Evolutionary Computation 2

Advanced Algorithms and Operators

Evolutionary Computation 2

Advanced Algorithms and Operators

Edited by

Thomas Bäck, David B Fogel
and Zbigniew Michalewicz

INSTITUTE OF PHYSICS PUBLISHING
Bristol and Philadelphia

INSTITUTE OF PHYSICS PUBLISHING
Bristol and Philadelphia

Copyright © 2000 by IOP Publishing Ltd

Published by Institute of Physics Publishing,
Dirac House, Temple Back, Bristol BS1 6BE, United Kingdom
(US Office: The Public Ledger Building, Suite 1035, 150 South Independence Mall West,
Philadelphia, PA 19106, USA)

British Library Cataloguing-in-Publication Data and
Library of Congress Cataloging-in-Publication Data are available

ISBN 0 7503 0665 3

PROJECT STAFF

Publisher: Nicki Dennis
Production Editor: Martin Beavis
Production Manager: Sharon Toop
Assistant Production Manager: Jenny Troyano
Production Controller: Sarah Plenty
Electronic Production Manager: Tony Cox

Printed in the United Kingdom

®™ The paper used in this publication meets the minimum requirements
of American National Standard for Information Sciences – Permanence of Paper
for Printed Library Materials, ANSI Z39.48-1984

Contents

Preface

The original *Handbook of Evolutionary Computation* (Bäck *et al* 1997) was designed to fulfill the need for a broad-based reference book reflecting the important role that evolutionary computation plays in a variety of disciplines—ranging from the natural sciences and engineering to evolutionary biology and computer sciences. The basic idea of evolutionary computation, which came onto the scene in the 1950s, has been to make use of the powerful process of natural evolution as a problem-solving paradigm, either by simulating it ('by hand' or automatically) in a laboratory, or by simulating it on a computer. As the history of evolutionary computation is the topic of one of the introductory sections of the *Handbook*, we will not go into the details here but simply mention that genetic algorithms, evolution strategies, and evolutionary programming are the three independently developed mainstream representatives of evolutionary computation techniques, and genetic programming and classifier systems are the most prominent derivative methods.

In the 1960s, visionary researchers developed these mainstream methods of evolutionary computation, namely J H Holland (1962) at Ann Arbor, Michigan, H J Bremermann (1962) at Berkeley, California, and A S Fraser (1957) at Canberra, Australia, for genetic algorithms, L J Fogel (1962) at San Diego, California, for evolutionary programming, and I Rechenberg (1965) and H P Schwefel (1965) at Berlin, Germany, for evolution strategies. The first generation of books on the topic of evolutionary compuation, written by several of the pioneers themselves, still gives an impressive demonstration of the capabilities of evolutionary algorithms, especially if one takes account of the limited hardware capacity available at that time (see Fogel *et al* (1966), Rechenberg (1973), Holland (1975), and Schwefel (1977)).

Similar in some ways to other early efforts towards imitating nature's powerful problem-solving tools, such as artificial neural networks and fuzzy systems, evolutionary algorithms also had to go through a long period of ignorance and rejection before receiving recognition. The great success that these methods have had, in extremely complex optimization problems from various disciplines, has facilitated the undeniable breakthrough of evolutionary computation as an accepted problem-solving methodology. This breakthrough is reflected by an exponentially growing number of publications in the field, and an increasing interest in corresponding conferences and journals. With these activities, the field now has its own archivable high-quality publications in which the actual research results are published. The publication of a considerable amount of application-specific work is, however, widely scattered over different disciplines and their specific conferences and journals, thus reflecting the general applicability and success of evolutionary computation methods.

The progress in the theory of evolutionary computation methods since 1990 impressively confirms the strengths of these algorithms as well as their limitations. Research in this field has reached maturity, concerning theoretical and application aspects, so it becomes important to provide a complete reference for practitioners, theorists, and teachers in a variety of disciplines. The original *Handbook of Evolutionary Computation* was designed to provide such a reference work. It included complete, clear, and accessible information, thoroughly describing state-of-the-art evolutionary computation research and application in a comprehensive style.

These new volumes, based in the original *Handbook*, but updated, are designed to provide the material in units suitable for coursework as well as for individual researchers. The first volume, *Evolutionary Computation 1: Basic Algorithms and Operators*, provides the basic information on evolutionary algorithms. In addition to covering all paradigms of evolutionary computation in detail and giving an overview of the rationale of evolutionary computation and of its biological background, this volume also offers an in-depth presentation of basic elements of evolutionary computation models according to the types of representations used for typical problem classes (e.g. binary, real-valued, permutations, finite-state machines, parse trees). Choosing this classification based on representation, the search operators mutation and recombination (and others) are straightforwardly grouped according to the semantics of the data they manipulate. The second volume, *Evolutionary Computation 2: Advanced Algorithms and Operators*, provides information on additional topics of major importance for the design of an evolutionary algorithm, such as the fitness evaluation, constraint-handling issues, and population structures (including all aspects of the parallelization of evolutionary algorithms). This volume also covers some advanced techniques (e.g. parameter control, meta-evolutionary approaches, coevolutionary algorithms, etc) and discusses the efficient implementation of evolutionary algorithms.

Organizational support provided by Institute of Physics Publishing makes it possible to prepare this second version of the *Handbook*. In particular, we would like to express our gratitude to our project editor, Robin Rees, who worked with us on editorial and organizational issues.

Thomas Bäck, David B Fogel and Zbigniew Michalewicz
August 1999

References

Bäck T, Fogel D B and Michalewicz Z 1997 *Handbook of Evolutionary Computation* (Bristol: Institute of Physics Publishing and New York: Oxford University Press)

Bezdek J C 1994 What is computational intelligence ? *Computational Intelligence: Imitating Life* ed J M Zurada, R J Marks II and C J Robinson (New York: IEEE Press) pp 1–12

Bremermann H J 1962 Optimization through evolution and recombination *Self-Organizing Systems* ed M C Yovits, G T Jacobi and G D Goldstine (Washington, DC: Spartan Book) pp 93–106

Fogel L J 1962 Autonomous automata *Industrial Research* **4** 14–9

Fogel L J, Owens A J and Walsh M J 1966 *Artificial Intelligence through Simulated Evolution* (New York: Wiley)

Fraser A S 1957 Simulation of genetic systems by automatic digital computers: I. Introduction *Austral. J. Biol. Sci.* **10** pp 484–91

Holland J H 1962 Outline for a logical theory of adaptive systems *J. ACM* **3** 297–314

——1975 *Adaptation in Natural and Artificial Systems* (Ann Arbor, MI: University of Michigan Press)

Rechenberg I 1965 Cybernetic solution path of an experimental problem *Royal Aircraft Establishment Library Translation No 1122* (Farnborough, UK)

Rechenberg I 1973 *Evolutionsstrategie: Optimierung technischer Systeme nach Prinzipien der biologischen Evolution* (Stuttgart: Frommann-Holzboog)

Schwefel H-P 1965 Kybernetische Evolution als Strategie der experimentellen Forschung in der Strömungstechnik *Diplomarbeit* Hermann Föttinger Institut für Strömungstechnik, Technische Universität, Berlin

——1977 *Numerische Optimierung von Computer-Modellen mittels der Evolutionsstrategie* Interdisciplinary Systems Research vol 26 (Basel: Birkhäuser)

List of Contributors

Peter J Angeline (Chapter 3)

Senior Scientist, Natural Selection, Inc., Vestal, NY, USA
e-mail: angeline@natural-selection.com

Thomas Bäck (Chapters 21, 18, Glossary)

*Associate Professor of Computer Science, Leiden University, The Netherlands; and
Managing Director and Senior Research Fellow, Center for Applied Systems
Analysis, Informatik Centrum Dortmund, Germany*
e-mail: baeck@ls11.informatik.uni-dortmund.de

James P Cohoon (Chapter 15)

*Associate Professor of Computer Science, University of Virginia, Charlottesville, VA,
USA*
e-mail: cohoon@virginia.edu

David W Coit (Chapter 7)

*Assistant Professor of Industrial Engineering, Rutgers University, Piscataway, NJ,
USA*
e-mail: coit@rci.rutgers.edu

Kalyanmoy Deb (Chapters 2, 14)

*Associate Professor of Mechanical Engineering, Indian Institute of Technology,
Kanpur, India*
e-mail: deb@iitk.ernet.in

A E Eiben (Chapters 20, 12)

*Leiden Institute of Advanced Computer Science, Leiden University, The Netherlands;
and Faculty of Sciences, Vrije Universiteit Amsterdam, The Netherlands*
e-mail: gusz@cs.leidenuniv.nl and gusz@cs.vu.nl

P J Fleming (Chapter 5)

Professor of Industrial Systems and Control, University of Sheffield, United Kingdom
e-mail: p.fleming@shef.ac.uk

David B Fogel (Glossary)

*Executive Vice President and Chief Scientist, Natural Selection Inc., La Jolla, CA,
USA*
e-mail: dfogel@natural-selection.com

C M Fonseca (Chapter 5)

*Postdoctoral Research Associate, Department of Automatic Control and Systems
Engineering, University of Sheffield, United Kingdom*
e-mail: c.fonseca@shef.ac.uk

Bernd Freisleben (Chapter 22)

Professor of Computer Science, University of Siegen, Germany
e-mail: freisleb@informatik.uni-siegen.de

John Grefenstette (Chapter 24)

*Head of the Machine Learning Section, Navy Center for Applied Research in
   Artificial Intelligence, Naval Research Laboratory, Washington, DC, USA*
e-mail: gref@aic.nrl.navy.mil

Tetsuya Higuchi (Chapter 26)

*Evolvable Hardware Systems Laboratory Leader, Electrotechnical Laboratory,
   Ibaraki, Japan*
e-mail: higuchi@etl.go.jp

Robert Hinterding (Chapter 20)

*Lecturer, Department of Computer and Mathematical Sciences, Victoria University of
   Technology, Melbourne, Australia*
rhh@matilda.vut.edu.au

Hitoshi Iba (Chapters 1, 4

Senior Researcher, Electrotechnical Laboratory, Ibaraki, Japan
e-mail: iba@etl.go.jp

Jens Lienig (Chapter 15)

Scientist, Physical Design Group, Tanner Research Inc., Pasadena, CA, USA
e-mail: jens.lienig@tanner.com

Samir W Mahfoud (Chapter 13)

*Vice President of Research and Software Engineering, Advanced Investment
   Technology, Clearwater, FL, USA*
e-mail: sam@ait-tech.com

Bernard Manderick (Chapter 26)

Professor of Computer Science, Free University Brussels, Belgium
e-mail: bernard@arti.vub.ac.be

W N Martin (Chapter 15)

*Associate Professor of Computer Science, University of Virginia, Charlottesville, VA,
   USA*
e-mail: martin@virginia.edu

Zbigniew Michalewicz (Chapters 8–11, 20)

*Professor of Computer Science, University of North Carolina, Charlotte, USA; and
   Institute of Computer Science, Polish Academy of Sciences, Warsaw, Poland*
e-mail: zbyszek@uncc.edu

Jan Paredis (Chapter 23)

Senoir Researcher, MATRIKS, University of Maastricht, The Netherlands
jan@riks.nl

Chrisila C Pettey (Chapter 16)

Assistant Professor of Computer Science, Middle Tennessee State University,
   Murfreesboro, TN, USA
e-mail: cscbp@knuth.mtsu.edu

Günter Rudolph (Chapter 25)

Senior Research Fellow, Center for Applied Systems Analysis, Informatik Centrum
   Dortmund, Germany
e-mail: rudolph@icd.de

Zs Ruttkay (Chapter 12)

Researcher, Faculty of Mathematics and Computer Science, Vrije Universiteit,
   Amsterdam, The Netherlands
e-mail: zsofi@cs.vu.nl

Alice E Smith (Chapter 7)

Professor, Department of Industrial and Systems Engineering, Auburn University,
   AL, USA
e-mail: aesmith@eng.auburn.edu

Robert E Smith (Chapter 17)

Senior Research Fellow, Intelligent Computing Systems Centre, Computer Studies
   and Mathematics Faculty, University of the West of England, Bristol, United
   Kingdom
rsmith@btc.uwe.ac.uk

William M Spears (Chapters 14, 19)

Commanding Officer, Naval Research Laboratory, Washington, DC, USA
e-mail: spears@aic.nrl.navy.mil

Jörg Ziegenhirt (Chapter 25)

Center for Applied Systems Analysis, Informatik Centrum Dortmund, Germany
e-mail: ziegenhi@ls11.informatik.uni-dortmund.de

Glossary

Thomas Bäck and David B Fogel

Bold text within definitions indicates terms that are also listed elsewhere in this glossary.

Adaptation: This denotes the general advantage in ecological or physiological efficiency of an **individual** in contrast to other members of the **population**, and it also denotes the process of attaining this state.

Adaptive behavior: The underlying mechanisms to allow living organisms, and, potentially, robots, to adapt and survive in uncertain environments (cf **adaptation**).

Adaptive surface: Possible biological trait combinations in a **population** of **individuals** define points in a high-dimensional sequence space, where each coordinate axis corresponds to one of these traits. An additional dimension characterizes the **fitness** values for each possible trait combination, resulting in a highly multimodal fitness landscape, the so-called adaptive surface or adaptive topography.

Allele: An alternative form of a **gene** that occurs at a specified chromosomal position (**locus**).

Artificial life: A terminology coined by C G Langton to denote the '... study of simple computer generated hypothetical life forms, i.e. life-as-it-could-be.' Artificial life and **evolutionary computation** have a close relationship because **evolutionary algorithms** are often used in artificial life research to breed the survival strategies of **individuals** in a **population** of artificial life forms.

Automatic programming: The task of finding a program which calculates a certain input–output function. This task has to be performed in automatic programming by another computer program (cf **genetic programming**).

Baldwin effect: Baldwin theorized that **individual** learning allows an organism to exploit genetic variations that only partially determine a physiological structure. Consequently, the ability to learn can guide evolutionary processes by rewarding partial genetic successes. Over evolutionary time, learning can guide evolution because individuals with useful genetic variations are maintained by learning, such that useful **genes** are utilized more widely in the subsequent generation. Over time, abilities that previously required learning are replaced by genetically determinant systems. The guiding effect of learning on evolution is referred to as the Baldwin effect.

Behavior: The response of an organism to the present environmental stimulus. The collection of behaviors of an organism defines the **fitness** of the organism to its present environment.

Boltzmann selection: The Boltzmann selection method transfers the probabilistic acceptance criterion of **simulated annealing** to **evolutionary algorithms**. The method operates by creating an offspring **individual** from two parents and accepting improvements (with respect to the parent's fitness) in any case and deteriorations according to an exponentially decreasing function of an exogeneous 'temperature' parameter.

Building block: Certain forms of recombination in **evolutionary algorithms** attempt to bring together building blocks, shorter pieces of an overall solution, in the hope that together these blocks will lead to increased performance.

Central dogma: The fact that, by means of **translation** and **transcription** processes, the genetic information is passed from the **genotype** to the **phenotype** (i.e. from **DNA** to **RNA** and to the **proteins**). The dogma implies that behaviorally acquired characteristics of an **individual** are not inherited to its offspring (cf **Lamarckism**).

Chromatids: The two identical parts of a duplicated **chromosome**.

Chromosome: Rod-shaped bodies in the nucleus of **eukaryotic cells**, which contain the hereditary units or **genes**.

Classifier systems: Dynamic, rule-based systems capable of learning by examples and induction. Classifier systems evolve a **population** of production rules (in the so-called Michigan approach, where an **individual** corresponds to a single rule) or a population of production rule bases (in the so-called Pittsburgh approach, where an individual represents a complete rule base) by means of an **evolutionary algorithm**. The rules are often encoded by a ternary alphabet, which contains a 'don't care' symbol facilitating a generalization capability of condition or action parts of a rule, thus allowing for an inductive learning of concepts. In the Michigan approach, the rule **fitness** (its strength) is incrementally updated at each generation by the 'bucket brigade' credit assignment algorithm based on the reward the system obtains from the environment, while in the Pittsburgh approach the fitness of a complete rule base can be calculated by testing the behavior of the individual within its environment.

Codon: A group of three nucleotide bases within the **DNA** that encodes a single amino acid or start and stop information for the **transcription** process.

Coevolutionary system: In coevolutionary systems, different **populations** interact with each other in a way such that the evaluation function of one population may depend on the state of the evolution process in the other population(s).

Comma strategy: The notation (μ, λ) strategy describes a selection method introduced in **evolution strategies** and indicates that a parent **population** of μ **individuals** generates $\lambda > \mu$ offspring and the best out of these λ offspring are deterministically selected as parents of the next generation.

Computational intelligence: The field of computational intelligence is currently seen to include subsymbolic approaches to artificial intelligence, such as **neural networks, fuzzy systems,** and **evolutionary computation,** which are gleaned from the model of information processing in natural systems. Following a commonly accepted characterization, a system is computationally intelligent if it deals only with numerical data, does not use knowledge in the classical expert system sense, and exhibits computational adaptivity, fault tolerance, and speed and error rates approaching human performance.

Convergence reliability: Informally, the convergence reliability of an **evolutionary algorithm** means its capability to yield reasonably good solutions in the case of highly multimodal topologies of the objective function. Mathematically, this is closely related to the property of global convergence with probability one, which states that, given infinite running time, the algorithm finds a global optimum point with probability one. From a theoretical point of view, this is an important property to justify the feasibility of evolutionary algorithms as **global optimization** methods.

Convergence velocity: In the theory of **evolutionary algorithms,** the convergence velocity is defined either as the expectation of the change of the distance towards the optimum between two subsequent generations, or as the expectation of the change of the objective function value between two subsequent generations. Typically, the best **individual** of a **population** is used to define the convergence velocity.

Crossover: A process of information exchange of genetic material that occurs between adjacent **chromatids** during **meiosis.**

Cultural algorithm: Cultural algorithms are special variants of **evolutionary algorithms** which support two models of inheritance, one at the microevolutionary level in terms of traits, and the other at the macroevolutionary level in terms of beliefs. The two models interact via a communication channel that enables the behavior of individuals to alter the belief structure and allows the belief structure to constrain the ways in which individuals can behave. The belief structure represents 'cultural' knowledge about a certain problem and therefore helps in solving the problem on the level of traits.

Cycle crossover: A **crossover** operator used in **order-based genetic algorithms** to manipulate permutations in a permutation preserving way. Cycle crossover performs recombination under the constraint that each element must come from one parent or the other by transferring element cycles between the mates. The cycle crossover operator preserves absolute positions of the elements of permutations.

Darwinism: The theory of evolution, proposed by Darwin, that evolution comes about through random variation (**mutation**) of heritable characteristics, coupled with **natural selection**, which favors those species for further survival and evolution that are best adapted to their environmental conditions.

Deception: Objective functions are called deceptive if the combination of good **building blocks** by means of **recombination** leads to a reduction of fitness rather than an increase.

Deficiency: A form of **mutation** that involves a terminal segment loss of **chromosome** regions.

Defining length: The defining length of a **schema** is the maximum distance between specified positions within the schema. The larger the defining length of a schema, the higher becomes its disruption probability by **crossover**.

Deletion: A form of **mutation** that involves an internal segment loss of a **chromosome** region.

Deme: An independent subpopulation in the **migration model** of parallel **evolutionary algorithms**.

Diffusion model: The diffusion model denotes a massively parallel implementation of **evolutionary algorithms**, where each **individual** is realized as a single process being connected to neighboring individuals, such that a spatial individual structure is assumed. **Recombination** and **selection** are restricted to the neighborhood of an individual, such that information is locally preserved and spreads only slowly over the **population**. (*See also Chapter 16.*)

Diploid: In diploid organisms, each body cell carries two sets of **chromosomes**; that is, each chromosome exists in two **homologous** forms, one of which is **phenotypically** realized.

Discrete recombination: Discrete recombination works on two vectors of **object variables** by performing an exchange of the corresponding object variables with probability one half (other settings of the exchange probability are in principle possible) (cf **uniform crossover**).

DNA: Deoxyribonucleic acid, a double-stranded macromolecule of helical structure (comparable to a spiral staircase). Both single strands are linear, unbranched nucleic acid molecules built up from alternating deoxyribose (sugar) and phosphate molecules. Each deoxyribose part is coupled to a nucleotide base, which is responsible for establishing the connection to the other strand of the DNA. The four nucleotide bases adenine (A), thymine (T), cytosine (C) and guanine (G) are the alphabet of the genetic information. The sequences of these bases in the DNA molecule determines the building plan of any organism.

Duplication: A form of **mutation** that involves the doubling of a certain region of a **chromosome** at the expense of a corresponding **deficiency** on the other of two **homologous** chromosomes.

Elitism: Elitism is a feature of some **evolutionary algorithms** ensuring that the maximum objective function value within a **population** can never reduce from one generation to the next. This can be assured by simply copying the best **individual** of a population to the next generation, if none of the selected offspring constitutes an improvement of the best value.

Eukaryotic cell: A cell with a membrane-enclosed nucleus and organelles found in animals, fungi, plants, and protists.

Evolutionary algorithm: *See* **evolutionary computation**.

Evolutionary computation: This encompasses methods of simulating evolution, most often on a computer. The field encompasses methods that comprise a population-based approach that relies on random variation and selection. Instances of algorithms that rely on evolutionary principles are called **evolutionary algorithms**. Certain historical subsets of evolutionary algorithms include **evolution strategies**, **evolutionary programming**, and **genetic algorithms**.

Evolutionary operation (EVOP): An industrial management technique presented by G E P Box in the late fifties, which provides a systematic way to test alternative production processes that result from small modifications of the standard parameter settings. From an abstract point of view, the method resembles a $(1 + \lambda)$ strategy with a typical setting of $\lambda = 4$ and $\lambda = 8$ (the so-called 2^2 and 2^3 factorial design), and can be interpreted as one of the earliest **evolutionary algorithms**.

Evolutionary programming: An **evolutionary algorithm** developed by L J Fogel at San Diego, CA, in the 1960s and further refined by D B Fogel and others in the 1990s. Evolutionary programming was originally developed as a method to evolve **finite-state machines** for solving time series prediction tasks and was later extended to parameter optimization problems. Evolutionary programming typically relies on variation operators that are tailored to the problem, and these often are based on a single parent; however, the earliest versions of evolutionary programming considered the possibility for recombining three or more **finite-state machines**. **Selection** is a stochastic **tournament selection** that determines μ **individuals** to survive out of the μ parents and the μ (or other number of) offspring generated by mutation. Evolutionary programming also uses the **self-adaptation** principle to evolve **strategy parameters** on-line during the search (cf **evolution strategy**).

Evolution strategy: An **evolutionary algorithm** developed by I Rechenberg and H-P Schwefel at the Technical University of Berlin in the 1960s. The evolution strategy typically employs real-valued parameters, though it has also been used for discrete problems. Its basic features are the distinction between a parent **population** (of size μ) and an offspring population (of size $\lambda \geq \mu$), the explicit emphasis on normally distributed **mutations**, the utilization of different forms of **recombination**, and the incorporation of the **self-adaptation** principle for **strategy parameters**; that is, those parameters that determine the mutation probability density function are evolved on-line, by the same principles which are used to evolve the **object variables**.

Exon: A region of **codons** within a **gene** that is expressed for the **phenotype** of an organism.

Finite-state machine: A transducer that can be stimulated by a finite alphabet of input symbols, responds in a finite alphabet of output symbols, and possesses some finite number of different internal states. The behavior of the finite-state machine is specified by the corresponding input–output symbol pairs and next-state transitions for each input symbol, taken over every state. In **evolutionary programming**, finite-state machines are historically the first structures that were evolved to find optimal predictors of the environmental behavior.

Fitness: The propensity of an **individual** to survive and reproduce in a particular environment. In **evolutionary algorithms**, the fitness value of an individual is closely related (and sometimes identical) to the objective function value of the solution represented by the individual, but especially when using **proportional selection** a **scaling function** is typically necessary to map objective function values to positive values such that the best-performing individual receives maximum fitness.

Fuzzy system: Fuzzy systems try to model the the fact that real-world circumstances are typically not precise but 'fuzzy'. This is achieved by generalizing the idea of a crisp membership function of sets by allowing for an arbitrary degree of membership in the unit interval. A fuzzy set is then described by such a generalized membership function. Based on membership functions, linguistic variables are defined that capture real-world concepts such as 'low temperature'. Fuzzy rule-based systems then allow for knowledge processing by means of fuzzification, fuzzy inference, and defuzzification operators which often enable a more realistic modeling of real-world situations than expert systems do.

Gamete: A **haploid** germ cell that fuses with another in fertilization to form a **zygote**.

Gene: A unit of **codons** on the **DNA** that encodes the synthesis for a **protein**.

Generation gap: The generation gap characterizes the percentage of the **population** to be replaced during each generation. The remainder of the population is chosen (at random) to survive intact. The generation gap allows for gradually shifting from the generation-based working scheme towards the extreme of just generating one new **individual** per 'generation', the so-called **steady-state selection** algorithm.

Genetic algorithm: An **evolutionary algorithm** developed by J H Holland and his students at Ann Arbor, MI, in the 1960s. Fundamentally equivalent procedures were also offered earlier by H J Bremermann at UC Berkeley and A S Fraser at the University of Canberra, Australia in the 1960s and 1950s. Originally, the genetic algorithm or adaptive plan was designed as a formal system for **adaptation** rather than an optimization system. Its basic features are the strong emphasis on **recombination** (crossover), use of a probabilistic **selection** operator (**proportional selection**), and the interpretation of **mutation** as a background operator, playing a minor role for the algorithm. While the original form of genetic algorithms (the canonical genetic algorithm) represents solutions by binary strings, a number of variants including real-coded genetic algorithms and order-based genetic algorithms have also been developed to make the algorithm applicable to other than binary search spaces.

Genetic code: The **translation** process performed by the ribosomes essentially maps triplets of nucleotide bases to single amino acids. This (redundant) mapping between the $4^3 = 64$ possible **codons** and the 20 amino acids is the so-called genetic code.

Genetic drift: A random decrease or increase of biological trait frequencies within the **gene** pool of a **population**.

Genetic programming: Derived from genetic algorithms, the genetic programming paradigm characterizes a class of **evolutionary algorithms** aiming at the automatic generation of computer programs. To achieve this, each **individual** of a **population** represents a complete computer program in a suitable programming language. Most commonly, symbolic expressions representing parse trees in (a subset of) the LISP language are used to represent these programs, but also other representations (including binary representation) and other programming languages (including machine code) are successfully employed.

Genome: The total genetic information of an organism.

Genotype: The sum of inherited characters maintained within the entire reproducing population. Often also the genetic constitution underlying a single trait or set of traits.

Global optimization: Given a function $f : M \to \mathbb{R}$, the problem of determining a point $x^* \in M$ such that $f(x^*)$ is minimal (i.e. $f(x^*) \le f(x) \; \forall x \in M$) is called the global optimization problem.

Global recombination: In **evolution strategies**, **recombination** operators are sometimes used which potentially might take all **individuals** of a **population** into account for the creation of an offspring individual. Such recombination operators are called global recombination (i.e. global **discrete recombination** or global **intermediate recombination**).

Gradient method: Local optimization algorithms for continuous parameter optimization problems that orient their choice of search directions according to the first partial derivatives of the objective function (its gradient) are called gradient strategies (cf **hillclimbing strategy**).

Gray code: A binary code for integer values which ensures that adjacent integers are encoded by binary strings with **Hamming distance** one. Gray codes play an important role in the application of canonical **genetic algorithms** to parameter optimization problems, because there are certain situations in which the use of Gray codes may improve the performance of an **evolutionary algorithm**.

Hamming distance: For two binary vectors, the Hamming distance is the number of different positions.

Haploid: Haploid organisms carry one set of genetic information.

Heterozygous: **Diploid** organisms having different **alleles** for a given trait.

Hillclimbing strategy: Hillclimbing methods owe their name to the analogy of their way of searching for a maximum with the intuitive way a sightless climber might feel his way from a valley up to the peak of a mountain by steadily moving upwards. These strategies follow a nondecreasing path to an optimum by a sequence of neighborhood moves. In the case of multimodal landscapes, hillclimbing locates the optimum closest to the starting point of its search.

Homologues: **Chromosomes** of identical structure, but with possibly different genetic information contents.

Homozygous: **Diploid** organisms having identical **alleles** for a given trait.

Hybrid method: **Evolutionary algorithms** are often combined with classical optimization techniques such as **gradient methods** to facilitate an efficient local search in the final stage of the evolutionary optimization. The resulting combinations of algorithms are often summarized by the term hybrid methods.

Implicit parallelism: The concept that each individual solution offers partial information about sampling from other solutions that contain similar subsections. Although it was once believed that maximizing implicit parallelism would increase the efficiency of an **evolutionary algorithm**, this notion has been proved false in several different mathematical developments (See **no-free-lunch theorem**).

Individual: A single member of a **population**. In **evolutionary algorithms**, an individual contains a **chromosome** or **genome**, that usually contains at least a representation of a possible solution to the problem being tackled (a single point in the search space). Other information such as certain **strategy parameters** and the individual's **fitness** value are usually also stored in each individual.

Intelligence: The definition of the term intelligence for the purpose of clarifying what the essential properties of artificial or computational intelligence should be turns out to be rather complicated. Rather than taking the usual anthropocentric view on this, we adopt a definition by D Fogel which states that intelligence is the capability of a system to adapt its behavior to meet its goals in a range of environments. This definition also implies that **evolutionary algorithms** provide one possible way to evolve intelligent systems.

Interactive evolution: The interactive evolution approach involves the human user of the **evolutionary algorithm** on-line into the variation–**selection** loop. By means of this method, subjective judgment relying on human intuition, esthetical values, or taste can be utilized for an evolutionary algorithm if a **fitness** criterion can not be defined explicitly. Furthermore, human problem knowledge can be utilized by interactive evolution to support the search process by preventing unnecessary, obvious detours from the global optimization goal.

Intermediate recombination: Intermediate recombination performs an averaging operation on the components of the two parent vectors.

Intron: A region of **codons** within a **gene** that do not bear genetic information that is expressed for the **phenotype** of an organism.

Inversion: A form of **mutation** that changes a **chromosome** by rotating an internal segment by 180° and refitting the segment into the chromosome.

Lamarckism: A theory of evolution which preceded Darwin's. Lamarck believed that acquired characteristics of an **individual** could be passed to its offspring. Although Lamarckian inheritance does not take place in nature, the idea has been usefully applied within some **evolutionary algorithms**.

Locus: A particular location on a **chromosome**.

Markov chain: A **Markov process** with a finite or countable finite number of states.

Markov process: A stochastic process (a family of random variables) such that the probability of the process being in a certain state at time k depends on the state at time $k - 1$, not on any states the process has passed earlier. Because the offspring **population** of an **evolutionary algorithm** typically depends only on the actual population, Markov processes are an appropriate mathematical tool for the analysis of evolutionary algorithms.

Meiosis: The process of cell division in **diploid** organisms through which germ cells (**gametes**) are created.

Metaevolution: The problem of finding optimal settings of the exogeneous parameters of an **evolutionary algorithm** can itself be interpreted as an optimization problem. Consequently, the attempt has been made to use an evolutionary algorithm on the higher level to evolve optimal **strategy parameter** settings for evolutionary algorithms, thus hopefully finding a best-performing parameter set that can be used for a variety of objective functions. The corresponding technique is often called a metaevolutionary algorithm. An alternative approach involves the **self-adaptation** of strategy parameters by evolutionary learning. (*See also Chapter 22.*)

Migration: The transfer of an **individual** from one subpopulation to another.

Migration model: The migration model (often also referred to as the island model) is one of the basic models of parallelism exploited by **evolutionary algorithm** implementations. The population is no longer **panmictic**, but distributed into several independent subpopulations (so-called **demes**), which coexist (typically on different processors, with one subpopulation per processor) and may mutually exchange information by interdeme **migration**. Each of the subpopulations corresponds to a conventional (i.e. sequential) evolutionary algorithm. Since **selection** takes place only locally inside a **population**, every deme is able to concentrate on different promising regions of the search space, such that the global search capabilities of migration models often exceed those of panmictic populations. The fundamental parameters introduced by the migration principle are the exchange frequency of information, the number of **individuals** to exchange, the selection strategy for the emigrants, and the replacement strategy for the immigrants. (*See also Chapter 15.*)

Monte Carlo algorithm: *See* **uniform random search**.

(μ, λ) strategy: *See* **comma strategy**.

($\mu + \lambda$) strategy: *See* **plus strategy**.

Multiarmed bandit: Classical analysis of **schema** processing relied on an analogy to sampling from a number of slot machines (one-armed bandits) in order to minimize expected losses.

Multimembered evolution strategy: All variants of **evolution strategies** that use a parent **population** size of $\mu > 1$ and therefore facilitate the utilization of **recombination** are summarized under the term multimembered evolution strategy.

Multiobjective optimization: In multiobjective optimization, the simultaneous optimization of several, possibly competing, objective functions is required. The family of solutions to a multiobjective optimization problem is composed of all those elements of the search space sharing the property that the corresponding objective vectors cannot be all simultaneously improved. These solutions are called Pareto optimal. (*See also Chapter 5.*)

Multipoint crossover: A **crossover** operator which uses a predefined number of uniformly distributed crossover points and exchanges alternating segments between pairs of crossover points between the parent **individuals** (cf **one-point crossover**).

Mutation: A change of the genetic material, either occurring in the germ path or in the **gametes** (generative) or in body cells (somatic). Only generative mutations affect the offspring. A typical classification of mutations distinguishes **gene** mutations (a particular gene is changed), **chromosome** mutations (the gene order is changed by **translocation** or **inversion**, or the chromosome number is changed by **deficiencies**, **deletions**, or **duplications**), and **genome** mutations (the number of chromosomes or genomes is changed). In **evolutionary algorithms**, mutations are either modeled on the **phenotypic** level (e.g. by using normally distributed variations with expectation zero for continuous traits) or on the **genotypic** level (e.g. by using bit inversions with small probability as an equivalent for nucleotide base changes).

Mutation rate: The probability of the occurrence of a **mutation** during **DNA** replication.

Natural selection: The result of competitive exclusion as organisms fill the available finite resource space.

Neural network: Artificial neural networks try to implement the data processing capabilities of brains on a computer. To achieve this (at least in a very simplified form regarding the number of processing units and their interconnectivity), simple units (corresponding to neurons) are arranged in a number of layers and allowed to communicate via weighted connections (corresponding to axons and dendrites). Working (at least principally) in parallel, each unit of the network typically calculates a weighted sum of its inputs, performs some internal mapping of the result, and eventually propagates a nonzero value to its output connection. Though the artificial models are strong simplifications of the natural model, impressive results have been achieved in a variety of application fields.

Niche: **Adaptation** of a **species** occurs with respect to any major kind of environment, the adaptive zone of this species. The set of possible environments that permit survival of a species is called its (ecological) niche.

Niching methods: In **evolutionary algorithms**, niching methods aim at the formation and maintenance of stable subpopulations (**niches**) within a single **population**. One typical way to achieve this proceeds by means of **fitness sharing** techniques. (*See also Chapter 13.*)

No-free-lunch theorem: This theorem proves that when applied across all possible problems, all algorithms that do not resample points from the search space perform exactly the same on average. This result implies that it is necessary to tune the operators of an **evolutionary algorithm** to the problem at hand in order to perform optimally, or even better than random search. The no-free-lunch theorem has been extended to apply to certain subsets of all possible problems. Related theorems have been developed indicating that

Object variables: The parameters that are directly involved in the calculation of the objective function value of an **individual**.

Off-line performance: A performance measure for **genetic algorithms**, giving the average of the best **fitness** values found in a **population** over the course of the search.

1/5 success rule: A theoretically derived rule for the deterministic adjustment of the standard deviation of the **mutation** operator in a $(1 + 1)$ **evolution strategy**. The 1/5 success rule reflects the theoretical result that, in order to maximize the **convergence velocity**, on average one out of five mutations should cause an improvement with respect to the objective function value.

One-point crossover: A **crossover** operator using exactly one crossover point on the **genome**.

On-line performance: A performance measure giving the average **fitness** over all tested search points over the course of the search.

Ontogenesis: The development of an organism from the fertilized **zygote** until its death.

Order: The order of a **schema** is given by the number of specified positions within the schema. The larger the order of a schema, the higher becomes its probability of disruption by **mutation**.

Order-based problems: A class of optimization problems that can be characterized by the search for an optimal permutation of specific items. Representative examples of this class are the traveling salesman problem or scheduling problems. In principle, any of the existing **evolutionary algorithms** can be reformulated for order-based problems, but the first permutation applications were handled by so-called order-based **genetic algorithms**, which typically use **mutation** and **recombination** operators that ensure that the result of the application of an operator to a permutation is again a permutation.

Order crossover: A **crossover** operator used in **order-based genetic algorithms** to manipulate permutations in a permutation preserving way. The order crossover (OX) starts in a way similar to **partially matched crossover** by picking two crossing sites uniformly at random along the permutations and mapping each string to constituents of the matching section of its mate. Then, however, order crossover uses a sliding motion to fill the holes left by transferring the mapped positions. This way, order crossover preserves the relative positions of elements within the permutation.

Order statistics: Given λ independent random variables with a common probability density function, their arrangement in nondecreasing order is called the order statistics of these random variables. The theory of order statistics provides many useful results regarding the moments (and other properties) of the members of the order statistics. In the theory of **evolutionary algorithms**, the order statistics are widely utilized to describe deterministic **selection** schemes such as the **comma strategy** and **tournament selection**.

Panmictic population: A mixed **population**, in which any **individual** may be mated with any other individual with a probability that depends only on **fitness**. Most conventional **evolutionary algorithms** have panmictic populations.

Parse tree: The syntactic structure of any program in computer programming languages can be represented by a so-called parse tree, where the internal nodes of the tree correspond to operators and leaves of the tree correspond to constants. Parse trees (or, equivalently, S-expressions) are the fundamental data structure in **genetic programming**, where **recombination** is usually implemented as a subtree exchange between two different parse trees.

Partially matched crossover: A **crossover** operator used to manipulate permutations in a permutation preserving way. The partially matched crossover (PMX) picks two crossing sites uniformly at random along the permutations, thus defining a matching section used to effect a cross through position-by-position exchange operations.

Penalty function: For constraint optimization problems, the penalty function method provides one possible way to try to achieve feasible solutions: the unconstrained objective function is extended by a penalty function that penalizes infeasible solutions and vanishes for feasible solutions. The penalty function is also typically graded in the sense that the closer a solution is to feasibility, the smaller is the value of the penalty term for that solution. By means of this property, an **evolutionary algorithm** is often able to approach the feasible region although initially all members of the population might be infeasible. (*See also Chapter 7.*)

Phenotype: The behavioral expression of the **genotype** in a specific environment.

Phylogeny: The evolutionary relationships among any group of organisms.

Pleiotropy: The influence of a single **gene** on several **phenotypic** features of an organism.

Plus strategy: The notation $(\mu + \lambda)$ strategy describes a **selection** method introduced in **evolution strategies** and indicates that a parent population of μ **individuals** generates $\lambda \geq \mu$ offspring and all $\mu + \lambda$ individuals compete directly, such that the μ best out of parents and offspring are deterministically selected as parents of the next generation.

Polygeny: The combined influence of several **genes** on a single **phenotypical** characteristic.

Population: A group of **individuals** that may interact with each other, for example, by mating and offspring production. The typical population sizes in **evolutionary algorithms** range from one (for $(1 + 1)$ **evolution strategies**) to several thousands (for **genetic programming**).

Prokaryotic cell: A cell lacking a membrane-enclosed nucleus and organelles.

Proportional selection: A **selection** mechanism that assigns selection probabilities in proportion to the relative **fitness** of an individual.

Protein: A multiply folded biological macromolecule consisting of a long chain of amino acids. The metabolic effects of proteins are basically caused by their three-dimensional folded structure (the tertiary structure) as well as their symmetrical structure components (secondary structure), which result from the amino acid order in the chain (primary structure).

Punctuated crossover: A **crossover** operator to explore the potential for **self-adaptation** of the number of crossover points and their positions. To achieve this, the vector of **object variables** is extended by a crossover mask, where a one bit indicates the position of a crossover point in the object variable part of the **individual**. The crossover mask itself is subject to recombination and mutation to allow for a self-adaptation of the crossover operator.

Rank-based selection: In rank-based selection methods, the selection probability of an **individual** does not depend on its absolute **fitness** as in case of **proportional selection**, but only on its relative fitness in comparison with the other population members: its rank when all individuals are ordered in increasing (or decreasing) order of fitness values.

Recombination: *See* **crossover**.

RNA: Ribonucleic acid.   The **transcription** process in the cell nucleus generates a copy of the nucleotide sequence on the coding strand of the **DNA**. The resulting copy is an RNA molecule, a single-stranded molecule which carries information by means of the necleotide bases adenine, cytosine, guanine, and uracil (U) (replacing the thymine in the DNA). The RNA molecule acts as a messenger that transfers information from the cell nucleus to the ribosomes, where the **protein** synthesis takes place.

Scaling function: A scaling function is often used when applying **proportional selection**, particularly when needing to treat individuals with non-positive evaluations. Scaling functions typically employ a linear, logarithmic, or exponential mapping.

Schema: A schema describes a subset of all binary vectors of fixed length that have similarities at certain positions. A schema is typically specified by a vector over the alphabet $\{0, 1, \#\}$, where the # denotes a 'wildcard' matching both zero and one.

Schema theorem: A theorem offered to describe the expected number of instances of a **schema** that are represented in the next generation of an **evolutionary algorithm** when **proportional selection** is used. Although once considered to be a 'fundamental' theorem, mathematical results show that the theorem does not hold in general when iterated over more than one generation and that it may not hold when individual solutions have noisy **fitness** evaluations. Furthermore, the theorem cannot be used to determine which schemata should be recombined in future generations and has little or no predictive power.

Segmented crossover: A **crossover** operator which works similarly to **multipoint crossover**, except that the number of crossover points is not fixed but may vary around an expectation value. This is achieved by a segment switch rate that specifies the probability that a segment will end at any point in the string.

Selection: The operator of **evolutionary algorithms**, modeled after the principle of **natural selection**, which is used to direct the search process towards better regions of the search space by giving preference to **individuals** of higher **fitness** for mating and reproduction. The most widely used selection methods include the **comma** and **plus strategies**, **ranking selection**, **proportional selection**, and **tournament selection**.

Self-adaptation: The principle of self-adaptation facilitates evolutionary algorithms learning their own **strategy parameters** on-line during the search, without any deterministic exogeneous control, by means of evolutionary processes in the same way as the **object variables** are modified. More precisely, the strategy parameters (such as **mutation** rates, variances, or covariances of normally distributed variations) are part of the individual and undergo mutation (**recombination**) and **selection** as the object variables do. The biological analogy consists in the fact that some portions of the **DNA** code for mutator **genes** or repair enzymes; that is, some partial control over the **DNA's** mutation rate is encoded in the **DNA**. (*See also Chapter 21.*)

Sharing: Sharing (short for **fitness** sharing) is a **niching method** that derates the fitnesses of **population** elements according to the number of **individuals** in a **niche**, so that the population ends up distributed across multiple niches.

Simulated annealing: An optimization strategy gleaned from the model of thermodynamic evolution, modeling an annealing process in order to reach a state of minimal energy (where energy is the analogue of **fitness** in **evolutionary algorithms**). The strategy works with one trial solution and generates a new solution by means of a variation (or **mutation**) operator. The new solution is always accepted if it represents a decrease of energy, and it is also accepted with a certain parameter-controlled probability if it represents an increase of energy. The control parameter (or **strategy parameter**) is commonly called temperature and makes the thermodynamic origin of the strategy obvious.

Speciation: The process whereby a new **species** comes about. The most common cause of speciation is that of geographical isolation. If a subpopulation of a single species is separated geographically from the main **population** for a sufficiently long time, its **genes** will diverge (either due to differences in **selection** pressures in different locations, or simply due to **genetic drift**). Eventually, genetic differences will be so great that members of the subpopulation must be considered as belonging to a different (and new) species.

Species: A **population** of similarly constructed organisms, capable of producing fertile offspring. Members of one species occupy the same ecological **niche**.

Steady-state selection: A **selection** scheme which does not use a generation-wise replacement of the **population**, but rather replaces one **individual** per iteration of the main **recombine–mutate–select** loop of the algorithm. Usually, the worst population member is replaced by the result of recombination and mutation, if the resulting individual represents a **fitness** improvement compared to the worst population member. The mechanism corresponds to a $(\mu + 1)$ selection method in **evolution strategies** (cf **plus strategy**).

Strategy parameter: The control parameters of an **evolutionary algorithm** are often referred to as strategy parameters. The particular setting of strategy parameters is often critical to gain good performance of an evolutionary algorithm, and the usual technique of empirically searching for an appropriate set of parameters is not generally satisfying. Alternatively, some researchers try techniques of **metaevolution** to optimize the strategy parameters, while in **evolution strategies** and **evolutionary programming** the technique of **self-adaptation** is successfully used to evolve strategy parameters in the same sense as **object variables** are evolved.

Takeover time: A characteristic value to measure the selective pressure of **selection** methods utilized in **evolutionary algorithms**. It gives the expected number of generations until, under repeated application of selection as the only operator acting on a **population**, the population is completely filled with copies of the initially best **individual**. The smaller the takeover time of a selection mechanism, the higher is its emphasis on reproduction of the best individual, i.e. its selective pressure.

Tournament selection: Tournament selection methods share the principle of holding tournaments between a number of **individuals** and selecting the best member of a tournament group for survival to the next generation. The tournament members are typically chosen uniformly at random, and the tournament sizes (number of individuals involved per tournament) are typically small, ranging from two to ten individuals. The tournament process is repeated μ times in order to select a population of μ members.

Transcription: The process of synthesis of a messenger **RNA** (mRNA) reflecting the structure of a part of the **DNA**. The synthesis is performed in the cell nucleus.

Translation: The process of synthesis of a **protein** as a sequence of amino acids according to the information contained in the messenger **RNA** and the **genetic code** between triplets of nucleotide bases and amino acids. The synthesis is performed by the ribosomes under utilization of transfer **RNA** molecules.

Two-membered evolution strategy: The two-membered or $(1+1)$ **evolution strategy** is an **evolutionary algorithm** working with just one ancestor individual. A descendant is created by means of **mutation**, and **selection** selects the better of ancestor and descendant to survive to the next generation (cf **plus strategy**).

Uniform crossover: A **crossover** operator which was originally defined to work on binary strings. The uniform crossover operator exchanges each bit with a certain probability between the two parent individuals. The exchange probability typically has a value of one half, but other settings are possible (cf **discrete recombination**).

Uniform random search: A random search algorithm which samples the search space by drawing points from a uniform distribution over the search space. In contrast to **evolutionary algorithms**, uniform random search does not update its sampling distribution according to the information gained from past samples, i.e. it is not a **Markov process**.

Zygote: A fertilized egg that is always **diploid**.

1

Introduction to fitness evaluation

Hitoshi Iba

1.1 Fitness evaluation

First, we describe how to encode and decode for fitness evaluations. Most
genetic algorithms require encoding; that is, the mapping from the chromosome
representation to the domain structures (e.g. parameters). The recombination
operators (i.e. crossover or mutation) work directly on this coded representation,
not on the domain structures. More formally, suppose that an optimization
problem is given as follows:

$$f : M \longrightarrow \Re \tag{1.1}$$

where M is the search space of the objective function f. Then the fitness
evaluation function F is described as follows:

$$F : R \xrightarrow{d} M \xrightarrow{f} \Re \xrightarrow{s} \Re_+ \tag{1.2}$$

$$F = s \circ f \circ d \tag{1.3}$$

where R is the space of the chromosome representation, d is a decoding
function, and s is a scaling function. The scaling function s is typically used
in combination with proportional selection in order to guarantee positive fitness
values and fitness maximization. For instance, when encoding an n-dimensional
real-valued objective function f_n by binary coding, the above fitness function
F is given as follows:

$$F : \{0, 1\}^l \xrightarrow{d_b} \Re^n \xrightarrow{f_n} \Re \xrightarrow{s} \Re_+ \tag{1.4}$$

where l is the length of a chromosome and d_b is the binary coding; that is,
d_b maps segments of the chromosome into real numbers of corresponding
dimensions. The evaluations of the chromosomes are converted into fitness
values in various ways. For instance, there are many coding schemes using a
binary character set that can code a parameter with the same meaning, such as
a binary code and a Gray code. However, experimental results have shown that

Gray coding is superior to binary coding for a particular function optimization for genetic algorithms (GAs) (see e.g. Caruana and Schaffer 1988). Analysis suggested that Gray coding eliminates the 'Hamming cliff' problem that makes some transitions difficult for a binary representation (see Bethke 1981, Caruana and Schaffer 1988, and Bäck 1993 for details). Therefore, the encoding scheme is very important to improve the search efficiency of GAs. The details are discussed in Chapter 2. In contrast to GAs, evolution strategies (ESs) and evolutionary *programming* work directly on the second space M, such that they do not require the decoding function d. Furthermore, they typically do not need a scaling function s, such that the fitness evaluation function is fully specified by equation (1.1).

1.2   Related problems

Second, it is often difficult to compute a solution with global accuracy for complex problems. The difficulty stems from the objectivity of the fitness function, which often comes only at the cost of significant knowledge about the search space (Angeline 1993). In order to eliminate the reliance on objective fitness functions, a competition is introduced. The competitive fitness function is a method for calculating fitness that is dependent on the current population, whereas the standard fitness functions return the same fitness for an individual regardless of other members in the population. The advantage of the competition is that evolutionary algorithms do not need an exact fitness value (i.e. the above f value), because most selection schemes work by just comparing fitness; that is the 'better or worse' criterion suffices. In other words, the absolute measure of fitness value is not required, but the relative measure when the individual is tested against other individuals should be derived. Thus, this method is computationally more efficient and more amenable to parallel implementation. The details are described in Chapter 3.

The third problem arises from the difficulty of evaluating the tradeoff between the fitness of a genotype and its complexity. For instance, the fitness definitions used in traditional genetic programming do not include evaluations of the tree descriptions. Therefore without the necessary control mechanisms, trees may grow exponentially large or become so small that they degrade search efficiency. Usually the maximum depth of trees is set as a user-defined parameter in order to control tree sizes, but an appropriate depth is not always known beforehand. For this purpose, we describe in Chapter 4 the complexity-based fitness evaluation by employing statistical measures, such as AIC and MDL.

Many real-world problems require a simultaneous optimization of multiple objectives. It is not necessarily easy to search for the different goals especially when they conflict with each other. Thus, there is a need for techniques different from the standard optimization in order to solve the multiobjective problem. Chapter 5 introduces several GA techniques studied recently.

References

Angeline P J 1993 *Evolutionary Algorithms and Emergent Intelligence* Doctoral Dissertation, Ohio State University

Bäck T 1993 Optimal mutation rates in genetic search *Proc. 5th Int. Conf. on Genetic Algorithms (Urbana-Champaign, IL, July 1993)* ed S Forrest (San Mateo, CA: Morgan Kaufmann) pp 2–9

Bethke A D 1981 *Genetic Algorithms as Function Optimizers* Doctoral Dissertation, University of Michigan

Caruana R A and Schaffer J D 1988 Representation and hidden bias: Gray vs binary coding for genetic algorithms *Proc. 5th Int. Conf. on Machine Learning (Ann Arbor, MI, 1988)* ed J Laird (San Mateo, CA: Morgan Kaufmann) pp 153–61

2

Encoding and decoding functions

Kalyanmoy Deb

2.1 Introduction

Among the EC algorithms, genetic algorithms (GAs) work with a coding of the object variables, instead of the variables directly. Thus, the encoding and decoding schemes are more relevant in the studies of GAs. Evolution strategy (ES) and *evolutionary programming* (EP) methods directly use the object variables. Thus, no coding is used in these methods. Genetic programming (GP) uses LISP codes to represent a task and no special coding scheme is usually used.

GAs begin their search by creating a population of solutions which are represented by a coding of the object variables. Before the fitness of each solution can be calculated, each solution must be decoded to obtain the object variables with a decoding function, $\gamma : \mathbb{B} \to \mathbb{M}$, where $\mathbb{M}$ represents a problem-specific space. Thus, the decoding functions are more useful from the GA implementation point of view, whereas the encoding functions (γ^{-1}) are important for understanding the coding aspects. The objective function $(f : \mathbb{M} \to \mathbb{R})$ in a problem can be calculated from the object variables defined in the problem-specific space $\mathbb{M}$. Thus the fitness function is a transformation defined as follows: $\Phi = f \circ \gamma$. It is important to mention here that both the above functions play an important role in the working of GAs. In addition, in the calculation of the fitness function, a scaling function is sometimes used after the calculation of f and γ functions. In the following subsections, we discuss different encoding and decoding schemes used in GA studies.

2.2 Binary strings

In most applications of GAs, binary strings are used to encode object variables. A binary string is defined using a binary alphabet $\{0, 1\}$. An l-bit string occupies a space $\mathbb{B}^l = \{0, 1\}^l$. Each object variable is encoded in a binary string of a particular length l_i defined by the user. Thereafter, a complete l-bit string is formed by concatenating all substrings together. Thus, the complete GA string has a length l:

$$l = \sum_{i=1}^{n} l_i \tag{2.1}$$

where n is the number of object variables. A binary string of length l_i has a total of 2^{l_i} search points. The string length used to encode a particular variable depends on the desired precision in that variable. A variable requiring a larger precision needs to be coded with a larger string and vice versa. A typical encoding of n object variables $x = (x_1, x_2, \ldots, x_n)$ into a binary string is illustrated in the following:

$$\underbrace{\overbrace{100\ldots1}^{x_1}}_{l_1} \quad \underbrace{\overbrace{010\ldots0}^{x_2}}_{l_2} \quad \cdots \quad \underbrace{\overbrace{110\ldots0}^{x_n}}_{l_n}.$$

The variable x_1 has a string length l_1 and so on. This encoding of the object variables to a binary string allows GAs to be applied to a wide variety of problems. This is because GAs work with the string and not with the object variables directly. The actual number of variables and the range of search domain of the variables used in the problem are masked by the coding. This allows the same GA code to be used in different problems without much change.

The decoding scheme used to extract the object variables from a complete string works in two steps. First, the substring $(a_{i1}, \ldots, a_{il_i})$, where $a_{ij} \in \mathbb{B}^{l_i}$, corresponding to each object variable is extracted from the complete string. Knowing the length of the substring and lower (u_i) and upper bounds (v_i) of the object variables, the following linear decoding function is mostly used $(\gamma_i : \mathbb{B}^{l_i} \to [u_i, v_i])$:

$$x_i = u_i + \frac{v_i - u_i}{2^{l_i} - 1} \left(\sum_{j=0}^{l_i-1} a_{i(l_i-j)} 2^j \right). \tag{2.2}$$

The above decoding function linearly maps the decoded integer value of the binary substring in the desired interval $[u_i, v_i]$. The above operation is carried out for all object variables. Thus, the operation $\gamma = \gamma_1 \times \ldots \times \gamma_n$ yields a vector of real values by interpreting the bit string as a concatenation of binary encoded integers mapped to the desired real space. As seen from the above decoding function, the maximum attainable precision in the ith object variable is $(v_i - u_i)/(2^{l_i} - 1)$. Knowing the desired precision and lower and upper bounds of each variable, a lower bound of the string size required to code the variable can be obtained.

Although the binary strings have been mostly used to encode object variables, higher-ary alphabets have also been used in some studies. In those cases, instead of a binary alphabet a higher-ary alphabet is used in a string. For a χ-ary alphabet string of length l, there are a total of χ^l strings possible.

Although the search space is larger with a higher-ary alphabet coding than with a binary coding of the same length, Goldberg (1990) has shown that the schema processing is maximum with binary alphabets. Nevertheless, in χ-ary alphabet coding the decoding function in equation (2.2) can be used by replacing 2 with χ.

In the above binary string decoding, the object variables are assumed to have a uniform search interval. For nonuniform but defined search intervals (such as exponentially distributed intervals and others), the above decoding function can be suitably modified. However, in some real-world search and optimization problems, the allowable values of the object variables do not usually follow any pattern. In such cases, the binary coding can be used, but the corresponding decoding function becomes cumbersome. In such cases, a look-up table relating a string and the corresponding value of the object variable is usually used (Deb and Goyal 1996).

Often in search and optimization problems, some object variables are allowed to take both negative and positive values. If the search interval in those variables is the same in negative and positive real space ($x_i \in \{-u_i, u_i\}$), a special encoding scheme is sometimes used. The first bit can be used to encode the sign of the variables and the other ($l_i - 1$) bits can be used to encode the magnitude of the variable (searching in the range $\{0, u_i\}$). It turns out that this encoding scheme is not very different from the simple binary encoding scheme applied over the entire search space.

2.3 Gray coded strings

Often, a *Gray coding* with binary alphabets is used to encode object variables (Caruna and Schaffer 1988, Schaffer *et al* 1989). Like the binary string, a Gray-coded string is also collection of binary alphabets of 1s and 0s. But the encoding and decoding schemes to obtain object variables from Gray-coded strings and vice versa are different. The encoding of the object variables to a Gray-coded string works in two steps. From the object variables, a corresponding binary string needs to be created. Thereafter, the binary string can be converted into a corresponding Gray code. A binary string $(b_1, b_2, \ldots, b_l)$, where $b_i \in \{0, 1\}$, is converted to a Gray code $(a_1, a_2, \ldots, a_l)$ by using a mapping $\gamma^{-1} : \mathbb{B}^l \to \mathbb{B}^l$ (Bäck 1993):

$$a_i = \begin{cases} b_i & \text{if } i = 1 \\ b_{i-1} \oplus b_i & \text{otherwise} \end{cases} \tag{2.3}$$

where $\oplus$ denotes addition modulo 2. As many researchers have indicated, the main advantage of a Gray code is its representation of adjacent integers by binary strings of Hamming distance one.

The decoding of a Gray-coded string into the corresponding object variables also works in two steps. First, the Gray-coded string $(a_1, \ldots, a_l)$ is converted into a simple binary string $(b_1, \ldots, b_l)$ as follows:

$$b_i = \bigoplus_{j=1}^{i} a_j \qquad \text{for } i = \{1, \ldots, l\}. \tag{2.4}$$

Thereafter, a decoding function similar to the one described in section C4.2.2 can be used to decode the binary string into a real number in the desired range $[u_i, v_i]$.

2.4 Messy coding

In the above coding schemes, the position of each gene is fixed along the string and only the corresponding bit value is specified. For example, in the binary string (101), the first and third genes take a value 1 and the second gene takes the value 0. If in a problem, a particular bit combination for some widely separated genes constitute a building block, it will be difficult to maintain the building block in the population under the action of the recombination operator. This problem is largely known as the *linkage* problem in GAs (Although a natural choice to bring the right gene combinations together is to use an inversion operator, Goldberg and Lingle (1985) have argued that inversion does not have an adequate search power to do the task in a reasonable time.) In order to solve the linkage problem, a different encoding scheme is suggested by Goldberg *et al* (1989). Both the gene position and the corresponding bit values are coded in a string. A typical four-bit string is coded as follows:

$$((2\ 1)\ (4\ 0)\ (1\ 1)\ (3\ 1))$$

The first entry inside a parenthesis is the gene location and the second entry is the bit value for that position. In the above string, the second, first and third genes have a value 1 and the fourth gene has a value 0. Thus, the above string is a representation of the binary string 1110. Since the gene location is also coded, good and important gene combinations can be expressed *tightly* (that is, adjacent to each other). This will reduce the chance of disruption of important building blocks due to the recombination operator. For example, if the bit-combination of 1 at the first gene and 0 at the fourth gene constitute a building block to the underlying problem, the above string codes the building block adjacent to each other. Thus, it will have a lesser chance of disruption due to the action of the recombination operator. This encoding scheme has been used to solve deceptive problems of various complexity (Goldberg *et al* 1989, Goldberg *et al* 1990).

2.5 Floating-point coding

Inspired by the success of the above flexible encoding scheme, Deb (1991), with assistance from Goldberg, has developed a floating-point encoding scheme for continuous variables. In that scheme, both mantissa and exponent of a floating-point parameter are represented by separate genes designated by M and E, respectively. For a multiparameter optimization problem, a typical gene has three elements, as opposed to two in the above encoding. The three elements are the parameter identification number, mantissa or exponent declaration, and its value. A typical two-parameter string is shown in the following:

((1 E +) (1 M 1) (1 E -) (2 M 1) (1 M 0) (1 M +) (2 E -) (2 E -) (1 M 0))

The decoding is achieved by first extracting mantissa and exponent values of each variable and then the parameter value is calculated using the following decoding function:

$$x_i = \text{mantissa}_i \times \text{base}^{\text{exponent}_i} \qquad (2.5)$$

where base is a fixed number (a value of 10 is suggested). A + and a - in the exponent gene indicate $+1$ and -1, respectively. In decoding the exponent of a parameter, first the number of + and - genes are counted. The exponent is then calculated by algebraically summing the number of + and - in the exponent genes. In the above string, the exponent in the first variable has one + and one -. Thus, the net exponent value is zero. For the second variable, there are no + and two -. Thus, the exponent value is -2. In order to decode the mantissa part of each variable, sets of 1 and 0 separated by either a + or a - are first identified in a left-to-right scan. Each set is then decoded by adaptively reducing interval depending on the length of each set. In the above string, the first variable has mantissa elements (in a left-to-right scan) 10+0. There are two sets of 1s and 0s that are separated by a +. In the first set, there are two bits with one 1 and one 0. With two bits, there are a total of three unary combinations possible: no 1, one 1, and two 1s. Dividing the mantissa search interval (0,1) into three intervals and denoting the first interval (0, 0.333) by no 1, the second interval (0.333, 0.667) by one 1, and the third interval (0.667, 1) by two 1s, we observe that the specified interval is the second interval. A + indicates that the decoded value of the next set has to be added to the lower limit of the current interval. Since in the next set there is only one bit, the current interval (0.333, 0.667) is now divided into two equal subintervals. Since the bit is a 0, we are in the first subinterval (0.333, 0.500). The decoded value of the mantissa of the first variable can then be taken as the average of the two final limits. As more mantissa bits are added to the right of the above string, the corresponding interval is continuously reduced and the accuracy of the solution is improved. Thus, the decoded value of the first parameter is $0.416(10^0)$ or 0.416, and that of the second parameter is $0.75(10^{-2})$ or 0.0075. This flexibility in coding allows important mantissa–exponent combinations to be coded tightly, thereby reducing the chance of disruption of good building blocks. Deb (1991) has used this encoding–decoding scheme to solve a difficult optimization problem and an engineering design problem.

2.6   Coding for binary variables

In some search problems, some of the object variables are binary denoting the presence or absence of a member. In network design problems, the presence of absence of a link in the network is often a object variable. In truss-structure design problems such as bridges and roofs, the presence or absence of a member is a design variable. In a neural network design problem, the presence or absence of a connection between two neurons is a decision variable. In these problems, the use of binary alphabets (1 for presence and 0 for absence) is most appropriate.

2.7   Coding for permutation problems

In permutation problems, such as the traveling salesperson problem and *scheduling* and *planning* problems, usually a series of node numbers is used to encode a permutation (Starkweather *et al* 1991). Usually in such problems a valid permutation requires each node number to appear once and only once. In these problems, the relative positioning of the node numbers are more important than the absolute positioning of the node numbers (Goldberg 1989). Although the sequence of node numbers to represent a permutation makes the encoding and decoding simpler, a few other problem-specific coding schemes have also been used in permutation problems (Whitley 1989).

2.8   Coding for control problems

In an optimal control problem, the decision variable is a time- or frequency-dependent function of some control variables. In applications of EC methods to optimal control problems, the practice has been to discretize the total time or frequency domain into several intervals and use the value of the control parameter at the beginning of each interval as a vector of object variables. In the case of a time-dependent control function $c(t)$ (from $t = t_1$ to $t = t_n$), the object variable vector ($x = \{x_1, \ldots . x_n\}$) is defined as follows ($\gamma^{-1} : \mathbb{R} \to \mathbb{R}$):

$$x_i = c(t = t_i). \tag{2.6}$$

In order to decode a vector of object variables into a time- or frequency-dependent control function, piece-wise spline approximation functions with adjacent object variables can be used (Goldberg 1989). In many optimal control problems, the control variable either monotonically increases or monotonically decreases with respect to the state variable (time of frequency). In such cases, an efficient coding scheme would be to use the difference in the control parameter values in two adjacent states, instead of the absolute value, as the object variable. For example, in the case of a monotonically increasing control variable, the following object variables can be used ($\gamma^{-1} : \mathbb{R} \to \mathbb{R}$):

$$x_i = \begin{cases} c(t_1) & i = 1 \\ c(t_i) - c(t_{i-1}) & \text{otherwise.} \end{cases} \tag{2.7}$$

The decoding function in this representation is little different than before. The control function can be formed by spline-fitting the adjacent control parameter values, which are calculated as follows ($\gamma : \mathbb{R} \to \mathbb{R}$):

$$c_i = \begin{cases} x_1 & i = 1 \\ x_i - x_{i-1} & \text{otherwise.} \end{cases} \tag{2.8}$$

2.9 Conclusions

Genetic algorithms, among other evolutionary computation methods, work mostly with a coding of object variables. The mapping of the object variables to a string code is achieved through an *encoding* function and the mapping of a string code to its corresponding object variable is achieved through a *decoding* function. In a binary coding, each object variable is discretized to take a finite number of values in a specified range. Although the binary coding has been popular, other coding schemes are also described, such as Gray coding to achieve a better variational property between encodings and corresponding decoded values, messy coding to achieve tight linkage of important gene combinations, and floating-point coding to have a unary coding scheme of real numbers. To take care of problems having binary object variables, permutation problems and optimal control problems using evolutionary computation algorithms, three different coding schemes are also discussed.

References

Bäck T 1993 Optimal mutation rates in genetic search *Proc. 5th Int. Conf. on Genetic Algorithms (Urbana, IL, July 1993)* ed S Forrest (San Mateo, CA: Morgan Kaufmann) pp 2–8

Caruana R A and Schaffer J D 1988 Representation and hidden bias: Gray versus binary coding in genetic algorithms *Proc. 5th Int. Conf. on Machine Learning (Ann Arbor, MI, 1988)* ed J Laird (San Mateo, CA: Morgan Kaufmann) pp 153–61

Deb K 1991 Binary and floating-point function optimization using messy genetic algorithms. (Doctoral dissertation, University of Alabama and IlliGAL Report No. 91004) *Dissertation Abstracts International* **52**(5), 2658B

Deb K and Goyal M 1996 A robust optimization procedure for mechanical component design based on genetic adaptive search *Technical Report No. IITK/ME/SMD-96001* Department of Mechanical Engineering, Indian Institute of Technology, Kanpur

Goldberg D E 1989 *Genetic Algorithms in Search, Optimization, and Machine Learning* (Reading, MA.: Addison-Wesley)

Goldberg D E 1990 Real-coded genetic algorithms, virtual alphabets, and blocking *IlliGal Report No 90001* (Urbana, IL: University of Illinois at Urbana-Champaign)

Goldberg D E, Deb K and Korb B 1990 Messy genetic algorithms revisited: Nonuniform size and scale *Complex Syst.* **4** 415–44

Goldberg D E, Korb B and Deb K 1989 Messy genetic algorithms: Motivation, analysis, and first results *Complex Syst.* **3** 493–530

Goldberg D E and Lingle R 1985 Alleles, loci, and the traveling salesman problem *Proc. 1st Int. Conf. on Genetic Algorithms (Pittsburgh, PA, July 1985)* ed J J Grefenstette (Hillsdale, NJ: Lawrence Erlbaum Associates) pp 154–9

Schaffer J D, Caruana R A, Eshelman L J, and Das R 1989 A study if control parameters affecting online performance of genetic algorithms for function optimization *Proc. 3rd Int. Conf. on Genetic Algorithms (Fairfax, VA, June 1989)* ed J D Schaffer (San Mateo, CA: Morgan Kaufmann) pp 51–60

Starkweather T, McDaniel S, Mathias K, Whitley D, and Whitley C 1991 A comparison of genetic sequencing operators *Proc. 4th Int. Conf. on Genetic Algorithms (San Diego, CA, July 1991)* ed R Belew and L Booker (San Mateo, CA: Morgan Kaufmann) pp 69–76

Whitley D, Starkweather T and Fuquay D 1989 Scheduling problems and traveling salesman: The genetic edge recombination operator *Proc. 3rd Int. Conf. on Genetic Algorithms (Fairfax, VA, June 1989)* ed J D Schaffer (San Mateo, CA: Morgan Kaufmann) pp 133–40

3

Competitive fitness evaluation

Peter J Angeline

3.1 Objective fitness

Typically in evolutionary computations, the value returned by the fitness function
is considered to be an exact, objective measure of the absolute worth of the
evaluated solution. More formally, the fitness function represents a complete
ordering of all possible solutions and returns a value for a given solution that is
related to its rank in the complete ordering. While in some environments such
'absolute objective' knowledge is easily obtained, it is often the case in real-
world environments that such information is inaccessible or unrepresentable.
Such fitness functions are sometimes called *objective fitness functions* (Angeline
and Pollack 1993).

Consider the following situation: given solutions A, B, and C, assume that A
is preferred to B, B is preferred to C, and C is preferred to A. Such a situation
occurs often in games where the C strategy can beat the A strategy, but is
beaten by the B strategy, which is in turn beaten by the A strategy. Note that
an objective fitness function cannot accurately represent such an arrangement.
If the objective fitness function gives A, B, and C the same fitness, then if
only A and B are in the population, the fact that A should be preferred to B is
unrepresented. The same holds for the other potential pairings. Similarly, if the
fitness function assigns distinct values for the worth of A, B, and C, there will
always be a pair of strategies that will have fitness values that do not reflect
their actual worth. The actual worth of any of the strategies A, B, or C, in this
case, is relative to the contents of the population. If all three are present then
none is to be preferred over the others; however, if only two are present then
there is a clear winner. In such problems, which are more prevalent than not,
no objective fitness function can adequately represent the space of solutions and
subsequently evolutionary computations will be misled.

3.2 Relative and competitive fitness

Relative fitness measures are an alternative to objective measures. Relative
fitness measures access a solution's worth through direct comparison to some

other solution either evolved or provided as a component of the environment. Competitive fitness is one type of relative fitness measure that is sensitive to the contents of the population. Sensitivity to the population is achieved through direct competition between population members. For such fitness measures, an objective fitness function that provides a partial ordering of the possible solutions is not required. All that is required is a relative measure of 'better' to determine which of two competing solutions is preferred.

The chief advantage of a competitive fitness functions are that they are self-scaling. Early in the run, when the evolving solutions perform poorly on the task, a solution need not be proficient in order to survive and reproduce. As the run continues and the average ability of the population increases, the average level of proficiency of a surviving solution will be suitably higher. As the population becomes increasingly better at solving the task, the difficulty of the fitness function is continually scaled commensurate with the ability of the average population members. Angeline and Pollack (1993) argue that competitive fitness measures set up an environment where complex ecologies of problem solving form that naturally encourage the emergence of generalized solutions.

Three types of competitive fitness function have been used in previous studies: full competition; bipartite competition; and a tournament competition. All of these are successful competitive fitness measures but they differ on the number of competitions required and necessity for additional objective measures.

Axelrod (1987) used a full competition where each player played every other player in the population, as is standard practice in game theory. The number of competitions required in such a scheme for a population of size N is

$$\frac{N(N-1)}{2}.$$

This is a considerable number of fitness evaluations, but it does provide a significant amount of information, and the number of competitions won by a player provides a sufficient amount of information for ranking the population members.

Hillis (1991) describes a genetic algorithm for evolving sorting networks using a bipartite competition scheme. In a bipartite competition, there are two teams (populations) and individuals from one team are played against individuals from the other. The total number of competitions played between population members is $N/2$ where N is the combined population size. While this method does provide significant feedback, it does not automatically produce a hierarchical ordering for the population. Consequently, an objective measure must be used to rank order the individuals at the completion of the competition for each of the two populations to determine which population members will reproduce.

Angeline and Pollack (1993) describe a single elimination tournament competitive fitness measure. In this method, each player is paired randomly with another player in the population with winners advancing to the next level of competition. Then all of the winners are again randomly paired and compete, with these winners again advancing to the next round. Play continues until a single individual is left, having beaten every competitor met in the tournament. This individual is designated as the best-of-run individual. The rank of the other population members is determined by the number of competitions they won before being eliminated. The number of competitions between N players in a single elimination tournament is in total $N - 1$, which is one fewer than the number of competitions held if each player played a user-designated teacher. Angeline and Pollack (1993) illustrate the presence of noise inherent in this method of fitness computation but claim that it may actually promote a more diverse population and ultimately help the evolutionary process.

The drawback of this fitness method is that it is often not clear how to create competitive fitness functions for problems that are not inherently competitive.

References

Angeline P J and Pollack J B 1993 Competitive environments evolve better solutions for complex tasks *Proc. 5th Int. Conf. on Genetic Algorithms (Urbana-Champaign, IL, July 1993)* ed S Forrest (San Mateo, CA: Morgan Kaufmann) pp 264–70

Axelrod R 1987 Evolution of strategies in the iterated prisoner's dilemma *Genetic Algorithms and Simulated Annealing* ed L Davis (Boston, MA: Pitman) pp 32–41

Hillis W D 1991 Co-evolving parasites improve simulated evolution as an optimization procedure *Emergent Computation* ed S Forrest (Cambridge, MA: MIT Press) pp 228–35

4

Complexity-based fitness evaluation

Hitoshi Iba

4.1 Introduction

Complexity-based fitness is grounded on a *simplicity criterion*, which is defined as a limitation on the complexity of the model class that may be instantiated when estimating a particular function. For example, when one is performing a polynomial fit, it seems fairly apparent that the degree of the polynomial must be less than the number of data points. Simplicity criteria have been studied by statisticians for many years.

This chapter outlines and compares the leading competing model selection criteria, namely, an MDL (minimum-description-length) principle, the AIC (Akaike information criterion), an MML (minimum-message-length) principle, the PLS (predictive least-squares) measure, cross-validation, and the maximum-entropy principle.

4.2 Model selection criteria

The complexity of an algorithm can be measured by the length of its minimal description in some language. The old but vague intuition of Occam's razor can be formulated as the *minimum-description-length criterion*; that is, given some data, the most probable model is the model that minimizes the sum (Weigend *et al* 1994):

$$
\begin{aligned}
\text{MDL(model)} \quad = \quad &\text{description\_length(data given model)} \\
&+ \text{description\_length(model)} \\
\longrightarrow \quad &\text{min} \qquad\qquad\qquad\qquad (4.1)
\end{aligned}
$$

where description_length(data given model) is the code length of the data when encoded using the model as a predictor for the data. The sum MDL(model) represents the tradeoff between residual error (i.e. the first term) and model complexity (i.e. the second term) including a structure estimation term for the final model. The final model (with the minimal MDL) is optimum in the sense

15

of being a consistent estimate of the number of parameters while achieving the minimum error (Tenorio and Lee 1990).

More formally, suppose that z_i is a sequence of observations from the random variable Z, which is characterized by probability function $p_Z(\theta)$. The dominant form of the MDL is

$$\text{MDL}(k) = -\log_2 p(z \mid \hat{\theta}) + \frac{k}{2} \log_2 N \qquad (4.2)$$

where $\hat{\theta}$ is the maximum-likelihood estimate of θ, $p(z \mid \hat{\theta})$ is the likelihood of the estimated density function of $p_Z(\theta)$, k is the number of parameters in the model, and N is the number of observations. The first term is the self-information of the model, which can be interpreted as the number of bits necessary to encode the observations. The second term can be also interpreted as the number of bits needed to encode the parameters of the model. Hence, the model which achieves the minimum of MDL is the most efficient model to encode the observations (Tenorio and Lee 1990, p 103).

Another criterion is the *Akaike information criterion* (AIC) (Akaike 1977). The essential idea here is to establish how many parameters, k, to include in a model. Minimizing AIC means minimizing k minus the log-likelihood function for the model, based on some assumed variance, $\hat{\sigma}^2$. In particular, if k is allowed to become too large, it does not matter that the likelihood of the data given a k-parameter model is very great; one will not achieve a minimal AIC. Unfortunately, the log-likelihood function cannot be calculated without an assumed family of distributions and a reasonable estimate of $\hat{\sigma}^2$. Nevertheless, the AIC has an important structural feature and that is the existence of a penalty term for the model complexity (Seshu 1994, p 220). The AIC is an approximation of the idealized Kullback–Leibler distance between the 'true' data generating distribution and the model, which involves the expectation operation.

Assuming the above condition of $\{z_1, \ldots, z_N\}$, the AIC estimator is given as

$$\text{AIC}(k) = -2 \ln p(z \mid \hat{\theta}) + 2k. \qquad (4.3)$$

By comparison with the MDL(k) criterion, we see that the difference is the crucial second term, k (AIC) versus $(k/2) \log_2 N$ (MDL). Therefore, if N is sufficiently large, the second term in equation 4.2 tends to penalize k much more severely for MDL than for AIC. The MDL(k) criterion penalizes the number of parameters asymptotically much more severely (Rissanen 1989, p 94). Moreover, under some conditions, it is assumed that learning generally converges much faster for MDL than for AIC (see the article by Yamanishi (1992) for details).

Wallace proposed a similar measure called MML (minimum message length). The coded form has two parts. The first states the inferred estimates of the unknown parameters in the model, and the second states the data using an optimal code based on the data probability distribution implied by these parameter estimates (Wallace and Freedman 1987). The total length might be interpreted as minus the log joint probability of estimate and data, and minimizing the length is therefore closely similar to maximizing the posterior probability of the estimate. MML is almost identical to MDL. However, they differ in the implementations and philosophical views as to prior probabilities. The details of these differences can be found in the article by Wallace and Freedman (1987) and its discussions.

The other criteria proposed are the *cross-validation* measure and the maximum-entropy principle. It is shown that qualitatively and asymptotically the cross-validation criterion is equivalent to AIC. We may consider that the maximum-entropy principle is a special case of the MDL principle, namely one where the model class is restricted to be of a special form. Within the statistical community, there is a considerable debate about both the proper viewpoint and the nature of the penalty term (Seshu 1994).

The goal shared by these complexity-based principles is to obtain accurate and parsimonious estimates of the probability distribution. The idea is to estimate the simplest density that has high likelihood by minimizing the total length of the description of the data. Barron introduced the index of resolvability, which may be interpreted as the MDL principle applied on the average. It is has been shown that the rate of convergence of minimum complexity estimators is bounded by the index of resolvability (Barron 1991).

Another useful criterion proposed is PLS (i.e. predictive least squares) or PSE (i.e. predicted square error) (Rissanen 1989, p 122). This is mostly aimed at solving the 'selection-of-variables' problem for linear regressions. The problem is solved by using the stochastic complexity and the sum of the prediction errors as a criterion, the latter either considered as an approximation of the former or as providing an independent extension of the LS principle. Rissanen described how to achieve the PLS solution to the posed regression problem, and revealed that the PLS criterion is a special case of the MDL principle. The detailed discussions are given by Rissanen (1989, chapter 5).

4.3 An example: minimum-description-length-based fitness evaluation for genetic programming

As an illustrative example, we present results of the experiments to evolve decision trees for Boolean concept learning using genetic programming (GP). We use the six-multiplexer problem as a means to treat the validity of MDL-based fitness functions. Decision trees were proposed by Quinlan for concept formation in machine learning (Quinlan 1983, 1986). Generating efficient decision trees from preclassified (supervised) training examples has generated a large literature. Decision trees can be used to represent Boolean concepts. Figure 4.1 shows a desirable decision tree which parsimoniously solves the six-multiplexer problem. In the six-multiplexer problem, a_0, and a_1 are the multiplexer addresses and d_0, d_1, d_2, and d_3 are the data. The target concept is

$$\text{output} = \overline{a_0}\,\overline{a_1}d_0 + a_0\overline{a_1}d_1 + \overline{a_0}a_1d_2 + a_0a_1d_3. \tag{4.4}$$

A decision tree is a representation of classification rules: for example, the subtree on the left in figure 4.1 shows that the output becomes false (i.e. zero) when the variable $a_0 = 0$, $a_1 = 0$, and $d_0 = 0$.

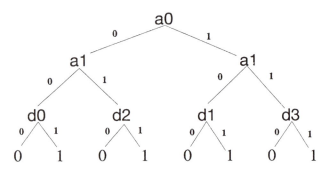

Figure 4.1. A more desirable and parsimonious decision tree.

Koza discussed the evolution of decision trees within a GP framework and conducted a small experiment called a *Saturday morning problem* (Koza 1990). However, Koza-style simple GP fails to evolve effective decision trees because an ordinary fitness function fails to consider parsimony. To overcome this shortcoming, we introduce fitness functions based on an MDL principle. This MDL-based fitness definition involves a tradeoff between the details of the tree, and the errors. In general, the MDL fitness definition for a GP tree (whose numerical value is represented by MDL) is defined as

$$\text{MDL} = (\text{exception\_coding\_length}) + (\text{tree\_coding\_length}) \tag{4.5}$$

where exception_coding_length is the description length of residual error (i.e. the first term in equation (4.1)) and tree_coding_length is the description length of the model.

The MDL value of a decision tree is calculated using the following method (Quinlan 1989). Consider the decision tree in figure 4.2 for the six-multiplexer problem (X, Y, and Z notations are explained later).

Using a breadth-first traversal, this tree is represented by the following string:

$$1\,d_0\,1\,a_1\,0\,0\,0\,0\,0\,1. \tag{4.6}$$

Since in our decision trees left (right) branches always represent zero (one) values for attribute-based tests, we can omit these attribute values. To encode this string in binary format,

$$2 + 3 + 2 \times \log_2 6 + 3 \times \log_2 2 = 13.17 \text{ bits} \tag{4.7}$$

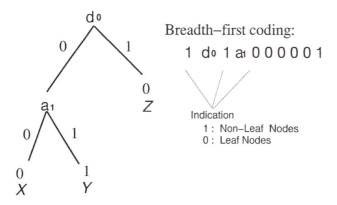

Figure 4.2. A decision tree for the six-multiplexer problem.

are required since the codes for each nonleaf (d_0, a_1) and for each leaf (0 or 1) require $\log_2 6$ and $\log_2 2$ bits respectively, and $2 + 3$ bits are used for their indications. In order to code exceptions (i.e. errors or incorrect classifications), their positions should be indicated. For this purpose, we divide the set of objects into classified subsets. For the tree of figure 4.2, we have three subsets (which we call X, Y, and Z from left to right) as shown in table 4.1. For instance, X is a subset whose members are classified into the leftmost leaf $(d_0 = 0 \wedge a_1 = 0)$. The number of elements belonging to X is 16. 12 members of X are correctly classified (i.e. 0). Misclassified elements (i.e. four elements of X, eight elements of Y, and 20 elements of Z) can be coded with the following cost:

$$L(16, 4, 16) + L(16, 8, 8) + L(32, 20, 32) = 65.45 \text{ bits} \quad (4.8)$$

where

$$L(n, k, b) = \log_2(b + 1) + \log_2 \left(\binom{n}{k} \right). \quad (4.9)$$

$L(n, k, b)$ is the total cost for transmitting a bitstring of length n, in which k of the symbols are ones and b is an upper bound on k. Thus the total cost for the decision tree in figure 4.2 is 78.62 ($=65.45 + 13.17$) bits.

In general, the coding length for a decision tree with n_f attribute nodes and n_t terminal nodes is given as follows:

$$\text{tree\_coding\_length} = (n_f + n_t) + n_t \log_2 T_s + n_f \log_2 F_s \quad (4.10)$$

$$\text{exception\_coding\_length} = \sum_{x \in \text{Terminals}} L(n_x, w_x, n_x) \quad (4.11)$$

where T_s and F_s are the total numbers of terminals and functions respectively. In equation (4.1), summation is taken over all terminal nodes. n_x is the number of elements belonging to the subset represented by x. w_x is the number of misclassified elements of n_x members.

Table 4.1. Classified subsets for encoding exceptions.

Name	Attributes	No of elements	No of correct classifications	No of incorrect classifications
X	$d_0 = 0 \wedge a_1 = 0$	16	12	4
Y	$d_0 = 0 \wedge a_1 = 1$	16	8	8
Z	$d_0 = 1$	32	12	20

Table 4.2. GP parameters.

Population size	100
Probability of graph crossover	0.6
Probability of graph mutation	0.0333
Terminal set	$\{0, 1\}$
Non-terminal set	$\{a_0, a_1, d_0, d_1, d_2, d_3\}$

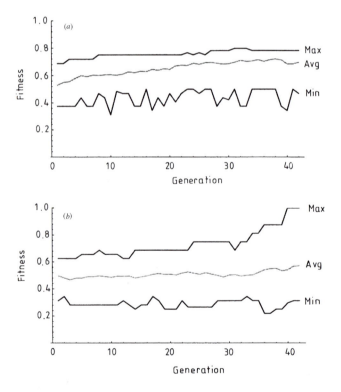

Figure 4.3. Experimental result: fitness versus generations, using (*a*) traditional (non-MDL) fitness function, and (*b*) MDL-based fitness function where the traditional (non-MDL) fitness is defined as the rate of correct classification.

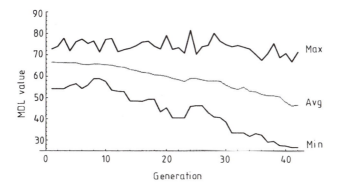

Figure 4.4. Experimental result: MDL versus generations.

With these preparations, we present results of the experiments to evolve decision trees for the six-multiplexer problem using GP. Table 4.2 shows the parameters used. A 1 (0) value in the terminal set represents a positive (negative) example, that is, a true (false) value. Symbols in the nonterminal set are attribute-based test functions. For the sake of explanation, we use S-expressions to represent decision trees from now on. The S-expression $(X \ Y \ Z)$ means that if X is 0 (false) then test the second argument Y and if X is 1 (true) then test the third argument Z. For instance, $(a_0 \ (d_1 \ (0) \ (1)) \ (1))$ is a decision tree which expresses that if a_0 is 0 (false) then if d_1 is 0 (false) then 0 (false) else 1 (true), and that if a_0 is 1 (true) then 1 (true).

Figure 4.3 shows results of experiments in terms of correct classification rate versus generations, using a traditional (non-MDL) fitness function (a), and using an MDL-based fitness function (b), where the traditional (non-MDL) fitness is defined as the rate of correct classification. The desired decision tree was acquired at the 40th generation when using an MDL-based fitness function. However, the largest fitness value (i.e. the rate of correct classification) at the 40th generation when using a non-MDL fitness function was only 78.12%. Figure 4.4 shows the evolution of the MDL values in the same experiment as figure 4.3(b). Figure 4.3(a) indicates clearly that the fitness test used in the non-MDL case is not appropriate for the problem. This certainly explains the lack of success in the non-MDL example. The acquired structure at the 40th generation using an MDL-based fitness function was as follows:

```
(A0
    (A1
        (D0
            (A1 (D0 (0) (0)) (D2 (0) (A0 (1) (1))))
            (1))
        (D2 (0) (1)))
    (A1 (D1 (0) (1)) (D3 (0) (1))))
```

whereas the typical genotype at the same generation (40th) using a non-MDL fitness function was as follows:

```
(D1
   (D2
      (D3 (0)
         (A0 (0)
            (D3 (A0 (0) (0))
               (D0 (D0 (1) (D1 (A0 (0) (1)) (D0 (D0 (0) (0)) (A1 (1) (1)))))
                  (D1 (0) (0))))))
      (A0
         (A0
            (D1 (1)
               (D1 (A0 (D1 (1) (1)) (A0 (D0 (D2 (1) (D1 (D1 (0) (0)) (1))) (0)) (1)))
                  (0)))
            (A0
               (A0
                  (D2
                     (A1 (1)
                        (D0
                           (D3 (0)
                              (A1 (D0 (A0 (1) (1)) (D3 (D0 (1) (0)) (A0 (0) (1))))
                                 (A1
                                    (D2
                                       (D2 (D0 (D2 (A1 (D3 (0) (D3 (0) (0))) (A1 (0) (1)))
                                                (D3 (D3 (0) (0)) (1)))
                                          (D2 (1) (D0 (1) (0))))
                                       (0))
                                    (D0 (D3 (D2 (A0 (D3 (0) (0)) (1)) (1)) (1)) (1))
                                    (1))))
                              (D0 (D3 (A0 (1) (A0 (0) (A1 (0) (A0 (1) (1))))) (D3 (1) (0))) (1))))
                        (1))
                     (A0 (A0 (0) (0)) (A1 (0) (D0 (1) (1)))))
                  (A1
                     (A0 (1)
                        (D3 (D1 (D2 (D1 (0) (A0 (1) (D3 (D1 (0) (1)) (1)))) (0)) (1)) (1))
                        (0))))
               (0)))
      (D2 (A1 (A1 (1) (0)) (A1 (1) (0))) (1))).
```

As can be seen, the non-MDL fitness function did not control the growth of the decision trees, whereas using an MDL-based fitness function led to a successful learning of a satisfactory data structure. Thus we can conclude that an MDL-based fitness function works well for the six-multiplexer problem.

4.4 Recent studies on complexity-based fitness

The complexity-based fitness evaluation can be introduced in order to control genetic algorithm (GA) search strategies. For instance, when applying GAs to the classification of genetic sequences, Konagaya and Konoto (1993) employed the MDL principle for GA fitness in order to avoid overlearning caused by the statistical fluctuations. They presented a GA-based methodology for learning stochastic motifs from given genetic sequences. A stochastic motif is a probabilistic mapping from a genetic sequence (which has been drawn from a finite alphabet) to a number of categories (such as cytochrome, globin, and trypsin). They employed Rissanen's MDL principle in selecting an optimal hypothesis (Yamanishi and Konagaya 1991).

When applying the MDL principle to GP, redundant structures should be pruned as much as possible, but at the same time premature convergence (i.e. premature loss of genotypic diversity) should be avoided (Zhang and Mühlenbein 1995). Zhang and Mühlenbein proposed a dynamic control to fix the error factor at each generation and change the complexity factor adaptively with respect to the error. Let $E_i(g)$ and $C_i(g)$ denote the error and complexity of the ith individual at generation g. Assuming that $0 \le E_i(g) \le 1$ and $C_i(g) \ge 0$, they defined the fitness of an individual i at generation g as

$$F_i(g) = E_i(g) + \alpha(g)C_i(g) \tag{4.12}$$

where $\alpha(g)$ is called the adaptive Occam factor and is expressed as

$$\alpha(g) = \begin{cases} \dfrac{1}{N^2}\dfrac{E_{\text{best}}(g-1)}{\check{C}_{\text{best}}(g)} & \text{if } E_{\text{best}}(g-1) > \epsilon \\[2ex] \dfrac{1}{N^2}\dfrac{1}{E_{\text{best}}(g-1)\check{C}_{\text{best}}(g)} & \text{otherwise} \end{cases} \tag{4.13}$$

where N is the size of the training set, E_{best} is the error value of the program which has the smallest (best) fitness value at generation $g-1$, and $\check{C}_{\text{best}}(g)$ is the size of the best program at generation g estimated at generation $g-1$, which is used for the normalization of the complexity factor. The user-defined constant ϵ specifies the maximum training error allowed for the final solution. Zhang and Mühlenbein have shown experimental results in the GP of sigma–pi neural networks. Their results were satisfactory.

In the articles by Iba and coworkers (1993, 1994), MDL-based fitness functions were applied successfully to system identification problems by using the implemented system STROGANOFF. The results showed that MDL-based fitness evaluation works well for tree structures in STROGANOFF, which controls GP-based tree search (see G1.7 for more details).

4.5 Conclusion

To conclude this section, we have shown that complexity-based fitness evaluations work by introducing a penalty term for the model complexity. We described several model selection criteria proposed so far. However the advantages and disadvantages of these approaches are not clear and are still being debated (see the article by Rissanen (1987) for details). Their different theoretical backgrounds and philosophies make it difficult to conduct comparative studies. The applicability and robustness of these methods remain to be seen as an interesting future topic.

References

Akaike H 1977 On the entropy maximization principle *Applications of Statistics* ed P R Krishnaiah (Amsterdam: North-Holland)

Barron A R 1991 Minimum complexity density estimation *IEEE Trans. Information Theory* **IT-37** 1034–54

Iba H, Kurita T, deGaris H and Sato T 1993 System identification using structured genetic algorithms *Proc. 5th Int. Conf. on Genetic Algorithms (Urbana-Champaign, IL, 1993)* ed S Forrest (San Mateo, CA: Morgan Kaufmann) pp 279–86

Iba H, deGaris H and Sato T 1994 Genetic programming using a minimum description length principle *Advances in Genetic Programming* ed K E Kinnear Jr (Cambridge, MA: MIT Press) pp 265–84

Konagaya A and Kondo H 1993 Stochastic motif extraction using a genetic algorithm with the MDL principle *Hawaii Int. Conf. on Computer Systems (1993)*

Koza J 1990 *Genetic Programming a Paradigm for Genetically Breeding Populations of Computer Programs to Solve Problems* Stanford University, Department of Computer Science, Report STAN-CS-90-1314

Quinlan J R 1983 Learning efficient classification procedures and their application to chess end games *Machine Learning* ed R S Michalski, J G Carbonell and T M Mitchell (Berlin: Springer) pp 463–82

——1986 Induction of decision trees *Machine Learning* **1** 81–106

——1989 Inferring decision trees using the minimum description length principles *Information Comput.* **80** 227–48

Rissanen J 1987 Stochastic complexity *J. R. Stat. Soc.* B **49** pp 223–39, 252–65

——1989 *Stochastic Complexity in Statistical Inquiry* (Singapore: World Scientific)

Seshu R 1994 Binary decision trees and an 'average-case' model for concept learning: implications for feature construction and the study of bias *Computational Learning Theory and Natural Learning Systems, Volume I: Constraints and Prospects* ed J Hanson *et al* (Cambridge, MA: MIT Press) pp 213–48

Tenorio M F and Lee W 1990 Self-organizing network for optimum supervised learning *IEEE Trans. Neural Networks* **NN-1** 100–9

Wallace C S and Freeman P R 1987 Estimation and inference by compact coding *J. R. Stat. Soc.* B **49** 240–65

Weigend A S and Rumelhart D E 1994 Weight elimination and effective network size *Computational Learning Theory and Natural Learning Systems, Volume I: Constraints and Prospects* ed J Hanson *et al* (Cambridge, MA: MIT Press) pp 457–76

Yamanishi K and Konagaya A 1991 Learning stochastic motifs from genetic sequences *Machine Learning (Proc. 8th Int. Workshop)* ed L A Birnbaum and G C Collins (San Mateo, CA: Morgan Kaufmann) pp 467–71

Yamanishi K A 1992 Learning criterion for stochastic rules *Machine Learning* (**9** 165–204

Zhang B-T and Mühlenbein H 1995 Balancing accuracy and parsimony in genetic programming *Evolutionary Comput.* **3** 17–38

5

Multiobjective optimization

C M Fonseca and P J Fleming

5.1 Introduction

Real-world problems often involve multiple measures of performance, or objectives, which should be optimized simultaneously. In practice, however, this is not always possible, as some of the objectives may be conflicting. Objectives are often also *noncommensurate*; that is, they measure fundamentally different aspects of the quality of a candidate solution. Thus, the quality of an individual is better described, not by a scalar, but by a vector quantity.

Performance, reliability, and cost are typical examples of conflicting, noncommensurate objectives. Improvement in any combination of these objectives will unequivocally improve the overall solution, but only as long as no degradation occurs in the remaining objectives. If this is not possible, then the current solution is said to be optimal in the Pareto sense, Pareto optimal, or nondominated. Otherwise, the new solution is said to dominate the old one. The set of all Pareto-optimal solutions is known as the Pareto-optimal set.

In most practical cases, a single compromise solution is sought. Thus, multiobjective optimization is generally more than purely searching for Pareto-optimal solutions. To be able to produce acceptable solutions, multiobjective optimization methods also need to take into account human preferences. In fact, although a Pareto-optimal solution should always be a better compromise solution than any solution it dominates, not all Pareto-optimal solutions may constitute acceptable compromise solutions.

5.2 Fitness evaluation

Multiobjective optimization with evolutionary algorithms, as with other optimizers, must ultimately be based on a scalarization of the objectives. Fitness, as a measure of the expected number of offspring of an individual, must remain a scalar. This scalarization should be a coordinatewise monotonic transformation, *but not necessarily a function*, of the objectives, so that individuals are always guaranteed to be at least as fit as those they dominate in the Pareto sense.

Such a transformation, being clearly nonunique, provides the necessary scope for incorporating preference information in the rating of the solutions. Once a scalar measure of quality (or cost) has been derived, the evolutionary algorithm may proceed with fitness assignment and selection as usual.

The cost assignment problem with multiple objectives is, essentially, a decision-making problem involving a finite number of objects, i.e. the individuals in the population, given what knowledge of the problem is available at the time of the decision. In this context, if a good decision strategy has been developed for a particular multiobjective problem, it should be possible to base the corresponding evaluation of fitness on that same strategy.

5.3 Current evolutionary approaches to multiobjective optimization

Current approaches to multiobjective optimization with evolutionary algorithms may be divided into three groups (Fonseca and Fleming 1995).

- *Plain aggregating approaches.* Objectives are numerically combined into a single objective *function* to be optimized.
- *Population-based non-Pareto approaches.* Different objectives affect the selection or deselection of different parts of the population in turn. Approaches based on separate rankings of the population according to each objective also fit in this category.
- *Pareto-based approaches.* The population is ranked making direct use of the definition of Pareto dominance.

Given the diversity of the approaches proposed in the literature to date, this classification is necessarily a rough one. However, it does reflect three of the main ideas behind the current handling of multiple objectives in evolutionary optimization, as the following review documents. Minimization of the objectives is assumed throughout except where noted otherwise.

5.3.1 The weighted-sum approach

Working mechanism. The n objectives $f_1, \ldots, f_n$ are weighted by user-defined, positive coefficients $w_1, \ldots, w_n$ and added together to obtain a scalar measure of cost for each individual. This measure can then be used as the basis for selection, e.g. proportional, *tournament*, or based on ranking. This approach is widely known, intuitive and simple to implement, and can be used with virtually all optimizers. Consequently, it is also the most popular.

Formal description.

$$\Phi : \mathbb{R}^n \longrightarrow \mathbb{R}$$
$$f(a_i) \longmapsto \sum_{k=1}^{n} w_k \, f_k(a_i)$$

where Φ denotes the cost assignment strategy.

Parameter settings. The setting of the weighting coefficients w_k is generally dependent on the problem instance, and not just on the problem class. Thus, the initial combination of weights usually needs to be fine tuned in order to lead to satisfactory compromise solutions. This usually implies rerunning the optimizer, although it may also be possible to modify the weights as the evolutionary algorithm runs. Hajela and Lin (1992) encoded the weights at the chromosome level, and promoted their diversity through *sharing*.

Theory. For any set of positive weights, the (global) optimum of Φ is always a nondominated solution of the original multiobjective problem (Goicoechea *et al* 1982). However, the opposite is not true. For example, nondominated solutions located in concave regions of the tradeoff surface cannot be obtained by this method, because their corresponding value of Φ is suboptimal (see, for example, the article by Fleming and Pashkevich (1985)). This is also illustrated in figure 5.1.

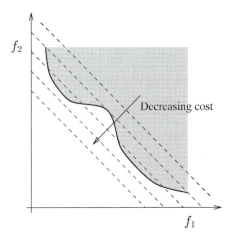

Figure 5.1. Lines of equal cost Φ induced by the weighted-sum approach (Fonseca 1995).

5.3.2 *The minimax approach*

Working mechanism. This approach consists of minimizing the maximum of the n objectives $f_1, \ldots, f_n$. In practice, it is often implemented as the minimization of the maximum (weighted) difference between the objectives and goals $g_1, \ldots, g_n$ specified by the user for each objective. Wilson and McLeod (1993) see this approach as a form of goal attainment (Gembicki 1974).

Formal description.

$$\Phi : \mathbb{R}^n \longrightarrow \mathbb{R}$$

$$\boldsymbol{f}(\boldsymbol{a}_i) \longmapsto \max_{k=1,\dots,n} \frac{f_k(\boldsymbol{a}_i) - g_k}{w_k}.$$

Parameter settings.   The goal values g_k indicate levels of performance in each objective dimension k to be approximated and, if possible, improved upon by the final solution.   In practice, the goals are often set to the desired levels of performance or, alternatively, to Utopian values known *a priori* to be unattainable.

The weights w_k indicate the desired direction of search in objective space, and are often set to the absolute value of the goals.   The smaller a weight, the *harder* the corresponding objective becomes with respect to the remaining objectives.   Hard objectives are essentially constraints, in that the corresponding goals must be attained, but only by a minimal amount.

Theory.   The minimax approach usually results in a cost function with regions of nonsmoothness, typically including the optimum, even if the objective functions themselves are smooth. For this reason, alternative formulations such as the goal attainment method (Gembicki 1974) are usually preferred to this approach when gradient-based optimizers are used. However, since evolutionary algorithms do not usually use gradient information, this should raise no concern.

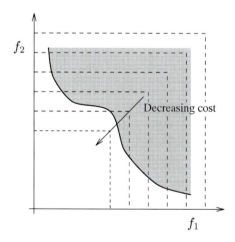

Figure 5.2. Lines of equal cost Φ induced by the minimax approach (Fonseca 1995).

Although this approach is able to produce solutions in concave regions of the tradeoff surface (see figure 5.2) the minimization of Φ is not guaranteed to produce strictly nondominated solutions. In fact, it is easy to show that one solution may dominate another, and yet have the same cost Φ.

5.3.3 The target vector approach

Working mechanism. This approach consists of minimizing the distance of the objective vector $\boldsymbol{f} = (f_1, \ldots, f_n)$ from a predefined goal, or target, vector $\boldsymbol{g} = (g_1, \ldots, g_n)$, according to a suitable distance measure (Wienke *et al* 1992).

Formal description.

$$\Phi : \mathbb{R}^n \longrightarrow \mathbb{R}$$
$$\boldsymbol{f}(\boldsymbol{a}_i) \longmapsto \left\| [\boldsymbol{f}(\boldsymbol{a}_i) - \boldsymbol{g}] \, \mathbf{W}^{-1} \right\|_\alpha .$$

Parameter settings. The goal values $g_1, \ldots, g_n$ indicate the desired levels of performance in each objective dimension, which are to be approximated by the final solution as closely as possible, typically in a weighted Euclidean sense ($\alpha = 2$). The weighting matrix $\mathbf{W}$ is often chosen to be diagonal, but in specific applications more elaborate weighting schemes may be appropriate (Wienke *et al* 1992).

5.3.4 The lexicographic approach

Working mechanism. Objectives are assigned distinct priorities according to how important they are, prior to optimization. The objective with the highest priority is used first when comparing individuals, either to decide a tournament (Fourman 1985) or to *rank* the population in a single-objective fashion. Any ties are resolved by comparing the relevant individuals again with respect to the second-highest-priority objective, and so forth, until the lowest-priority objective is reached.

Formal description.

$$\Phi : \mathbb{R}^n \longrightarrow \{0, 1, \ldots, \mu - 1\}$$
$$\boldsymbol{f}(\boldsymbol{a}_i) \longmapsto \sum_{j=1}^{\mu} \mathbb{1}\left(\boldsymbol{f}(\boldsymbol{a}_j) \, \ell< \boldsymbol{f}(\boldsymbol{a}_i)\right)$$

where $\boldsymbol{f}(\boldsymbol{a}_j) \, \ell< \boldsymbol{f}(\boldsymbol{a}_i)$ if and only if

$$\exists\, p \in \{1, \ldots, n\} : \forall k \in \{p, \ldots, n\} \qquad f_k(\boldsymbol{a}_j) \leq f_k(\boldsymbol{a}_i) \wedge f_p(\boldsymbol{a}_j) < f_p(\boldsymbol{a}_i)$$

and where $\mathbb{1}$ (condition) evaluates to unity if the condition is verified, and to zero otherwise.

The objectives $f_1, \ldots, f_n$ are assumed to be sorted in order of increasing priority.

Parameter settings. All objectives must be assigned distinct priorities. This requirement will be acceptable in those practical situations where decisions regarding the quality of a solution are made sequentially (Ben-Tal 1980), and where the relative importance of the various objectives is well understood.

In the case of heavily competing objectives, the final solution may vary wildly depending on how priorities are assigned. For example, if all objectives admit single, but different, optima, the lexicographic optimum will be no more than the optimum of the highest-priority objective.

Theory. Lexicographically optimal solutions are also, by definition, Pareto-optimal solutions.

5.3.5 The VEGA approach

Working mechanism. The vector-evaluated genetic algorithm (VEGA, Schaffer 1985) was probably the first evolutionary approach explicitly aimed at promoting the generation of multiple nondominated solutions. Subpopulations of offspring are selected according to each objective in turn, using *fitness-proportionate* selection. Offspring are then mixed in order to undergo recombination and mutation, regardless of which objective dictated their selection.

Formal description. The merging of the offspring subpopulations in VEGA is equivalent to averaging the fitness components δ_k corresponding to each objective f_k (here maximization is assumed).

$$\Delta : \mathbb{R}^n \longrightarrow \mathbb{R}$$

$$f(a_i) \longmapsto \sum_{k=1}^{n} \lambda_k \, \delta_k \left[f_k(a_i) \right]$$

where, since proportional selection is used,

$$\delta_k \left[f_k(a_i) \right] = \frac{f_k(a_i)}{\sum_{j=1}^{\mu} f_k(a_j)} \, .$$

Note that, whereas Φ has been used earlier to denote a cost assignment strategy on which selection can be based, Δ is used here to denote a multiobjective *fitness* assignment strategy, in analogy with the use of δ to denote a single-objective scaling function.

Parameter settings. The size of each subpopulation, λ_k, controls the involvement of the associated objective in the selection process. In Schaffer's original work (Schaffer 1985), all subpopulations were the same size.

Theory. On concave tradeoff surfaces, the population may split into *species* particularly strong in each objective (*speciation*, Schaffer 1985). This undesirable effect has been noted to arise from VEGA ultimately performing a weighted sum of the objectives (Richardson *et al* 1989), even though the weights associated with this linear combination depend on the distribution of the population at each generation.

However, by effectively weighting each objective proportionally to the inverse of the total population fitness in that objective dimension, VEGA adaptively attempts to balance improvement in the various objective dimensions. If improvement is observed only in some objectives, selection will then favor improvement in the remaining objectives. As a result, VEGA can, at least in some cases, maintain different species much longer than a GA optimizing a pure weighted-sum would do, due to genetic drift (Fonseca and Fleming 1995).

5.3.6 The median-rank approach

Working mechanism. The population is first ranked according to each single objective, separately. Then, the median of the ranks assigned to each individual is computed and used for fitness assignment (Breeden 1995).

Variations of this approach include implementations where the average is used to replace the median, and where fitness values are computed based on each ranking first, and only subsequently averaged. Implementations where objectives are used in turn to decide tournaments (Fourman 1985), or to dictate the deletion of fractions of the population (Kursawe 1991), can also be seen as fitting in this category.

Formal description.

$$\Phi : \mathbb{R}^n \longrightarrow [0, \mu - 1]$$
$$\boldsymbol{f}(\boldsymbol{a}_i) \longmapsto \text{median}\{\text{rank}_1[f_1(\boldsymbol{a}_i)], \ldots, \text{rank}_n[f_n(\boldsymbol{a}_i)]\}.$$

Parameter settings. The plain median-rank approach implicitly assumes that all objectives are equally important. The median analogue of a weighted sum can nevertheless be achieved by entering the rank value associated with each objective a number of times proportional to its importance, and computing the median of the resulting sample (Breeden 1995).

Computing the average instead of the median may offer a simpler way of controlling the relative importance of each objective, but the resulting algorithm will also be more sensitive to any uncertainty in the evaluation of the objective functions.

Theory. Ranking the population according to each objective separately avoids the normalization difficulties associated with all aggregating function approaches. As a consequence, and despite the similarity to the weighted-sum approach, algorithms based on these methods are not affected by whether tradeoff surfaces are convex or concave.

5.3.7 Pareto ranking

Working mechanism. The concept of Pareto dominance is used to rank the population in such a way that all nondominated individuals in the population are assigned the same cost. Two approaches to Pareto ranking have been proposed in the literature.

(i) In the approach originally proposed by Goldberg (1989), all nondominated individuals in the population are assigned a cost of one, and removed from contention. Then, a new set of nondominated individuals is identified and assigned a cost of two. The process continues until the whole population has been ranked.

(ii) An alternative approach has been proposed by Fonseca and Fleming (1993), where individuals are simply assigned a cost value according to how many individuals in the population they are dominated by.

Once computed, these cost values are used to perform selection, e.g. using rank-based fitness assignment (Fonseca and Fleming 1993, Srinivas and Deb 1994), or tournament selection (Cieniawski 1993, Ritzel *et al* 1994). Rather than ranking the population first, Horn *et al* (1994) based their tournament selection directly on whether or not one of the competitors dominated the other one.

Goldberg (1989) has also noted the need for niching (Chapter 13) and *speciation* (Chapter 14) techniques in order to stabilize the population along the Pareto front. Most of the implementations of Pareto-based fitness assignment cited above also include techniques such as fitness sharing (Section 13.2)and *mating restriction* (Section 14.4).

Formal description.

(i) The first ranking strategy will be defined through recursion:

$$\Phi : \mathbb{R}^n \longrightarrow \{1, 2, \ldots, \mu\}$$

$$f(a_i) \longmapsto \begin{cases} 1 & \Leftarrow \quad \neg[f(a_j) \; \mathrm{p}{<} \; f(a_i)] \\ & \qquad \forall j \in \{1, \ldots, \mu\} \\[1em] \phi & \Leftarrow \quad \neg[f(a_j) \; \mathrm{p}{<} \; f(a_i)] \\ & \qquad \forall j \in \{1, \ldots, \mu\} \setminus \{l : \Phi[f(a_l)] < \phi\} \end{cases}$$

where $f(a_j) \; \mathrm{p}{<} \; f(a_i)$ if and only if

$$\forall k \in \{1, \ldots, n\} \qquad f_k(a_j) \leq f_k(a_i) \wedge \exists k \in \{1, \ldots, n\} : f_k(a_j) < f_k(a_i)$$

and where the symbol $\neg$ denotes logical negation.

(ii) The second ranking strategy follows is simpler to define:

$$\Phi : \mathbb{R}^n \longrightarrow \{0, 1, \ldots, \mu - 1\}$$

$$f(a_i) \longmapsto \sum_{j=1}^{\mu} \mathbb{1}\left(f(a_j) \; p< \; f(a_i)\right)$$

where $\mathbb{1}$ (condition) evaluates to unity if the condition is verified and to zero otherwise.

Parameter settings. There are no parameters to set. This is especially attractive if the relative importance of the different objectives is not known *a priori*, or cannot be formally expressed easily. Unfortunately, this also means that, even if preference information is indeed available, it cannot be used.

Other than implementation issues, there is currently no reason to prefer one ranking strategy to the other. The second approach seems to be easier to interpret in the (theoretical) infinite-population case (Fonseca and Fleming 1995), and thus may be easier to analyze.

Theory. Both ranking procedures are such that $f(a_j) \; p< \; f(a_i)$ implies $\Phi[f(a_j)] < \Phi[f(a_i)]$, and that all nondominated individuals are assigned the same cost. As a consequence, it is possible for the population to stagnate if it enters a state where most individuals are nondominated. This is especially likely to happen as the number of competing objectives increases.

The ranks assigned by method (ii) to a large uniformly distributed population, once normalized by the population size, may be seen as estimates of what fraction of the search space outperforms each particular point considered. For problems involving two decision variables only, this interpretation of ranking allows the visualization of the cost landscapes induced by Pareto ranking, as well as by ranking based on other concepts, such as lexicographic optimality (discussed earlier), and *preferability* given a goal vector, which will be discussed next.

5.3.8 *Pareto-like ranking with goal and priority information*

Working mechanism. In this approach, a concept which combines Pareto dominance with goal and priority information is used instead of pure Pareto dominance to rank the population by the second method described above, making it possible to bias the search away from regions of the tradeoff surface known *a priori* to be unacceptable.

As in the lexicographic approach, higher-priority objectives come into play before those with a lower priority, but different objectives may now be assigned equal priorities. In addition to this, the comparison of solutions is affected by whether or not they attain the goals set for the various objectives.

Since the setting of goals and priorities ultimately depends on the personal preference of the operator, this concept has been called *preferability* (Fonseca 1995).

Formal description.

$$\Phi : \mathbb{R}^n \longrightarrow \{0, 1, \ldots, \mu - 1\}$$

$$f(a_i) \longmapsto \sum_{j=1}^{\mu} \mathbb{1}\left(f(a_j) \underset{g}{\prec} f(a_i) \right)$$

where the symbol $\underset{g}{\prec}$ indicates preferability given the *preference vector g*.

To define preferability, consider the n-dimensional preference vector g, where n is the number of objectives, as a concatenation of p vectors g_m, $m = 1, \ldots, p$. Each subvector g_m contains those n_m components of g which have been assigned priority m, such that

$$\sum_{m=1}^{p} n_m = n.$$

Also, let the smile $\overset{j}{\smile}$ index the components of f and g where $f(a_j) \le g$, and let the frown $\overset{j}{\frown}$ index the remaining components of these vectors, for each given individual a_j, and similarly for f_m and g_m, $m = 1, \ldots, p$. Then, $f(a_j) \underset{g}{\prec} f(a_i)$ if and only if

$$p = 1 \Rightarrow \left(f_p(a_j)^{\overset{j}{\frown}} \text{ p<} f_p(a_i)^{\overset{j}{\frown}} \right) \vee \left\{ \left(f_p(a_j)^{\overset{j}{\frown}} = f_p(a_i)^{\overset{j}{\frown}} \right) \right.$$
$$\left. \wedge \left[\left(f_p(a_i)^{\overset{j}{\smile}} \nleq g_p^{\overset{j}{\smile}} \right) \vee \left(f_p(a_j)^{\overset{j}{\smile}} \text{ p<} f_p(a_i)^{\overset{j}{\smile}} \right) \right] \right\}$$

and

$$p > 1 \Rightarrow \left(f_p(a_j)^{\overset{j}{\frown}} \text{ p<} f_p(a_i)^{\overset{j}{\frown}} \right) \vee \left\{ \left(f_p(a_j)^{\overset{j}{\frown}} = f_p(a_i)^{\overset{j}{\frown}} \right) \right.$$
$$\left. \wedge \left[\left(f_p(a_i)^{\overset{j}{\smile}} \nleq g_p^{\overset{j}{\smile}} \right) \vee \left(f_{1,\ldots,p-1}(a_j) \underset{g_{1,\ldots,p-1}}{\prec} f_{1,\ldots,p-1}(a_i) \right) \right] \right\}$$

where $f_{1,\ldots,p-1}$ denotes the concatenation of $f_1, \ldots, f_{p-1}$, and similarly for $g_{1,\ldots,p-1}$.

Parameter settings.   By setting goals and assigning priorities to the objectives, a number of other, simpler cost assignment strategies can be implemented (Fonseca 1995).

- *Pareto.* All objectives are assigned priority 1 and fully unattainable goals; that is,

$$g = g_1 = [(-\infty, \ldots, -\infty)].$$

- *Lexicographic.* Objectives are all assigned different priorities and fully unattainable goals; that is,

$$g = (g_1, \ldots, g_n) = [(-\infty), \ldots, (-\infty)].$$

- *Constrained optimization.* Inequality constraints are handled as priority 2 objectives to be minimized until the corresponding goals are reached. Assuming Pareto optimization for the soft objectives, one has

$$g = (g_1, g_2) = [(-\infty, \ldots, -\infty), (g_{2\,1}, \ldots, g_{2\,n_c})]$$

where n_c is the number of constraints. If there are no soft objectives, the problem becomes a constraint satisfaction problem:

$$g = (g_2) = [(g_{2,1}, \ldots, g_{2,n_c})].$$

- *Goal programming.* Goals are set as for the minimax approach, and all objectives are assigned priority 1:

$$g = (g_1) = [(g_{1,1}, \ldots, g_{1,n})].$$

These parameters would promote the sampling of the portion of the tradeoff surface which dominates the goals, if they are attainable, or, if they are unattainable, the region dominated by the goals.

Goals and priorities may also be set interactively during an optimization session (Fonseca and Fleming 1993, Fonseca 1995).

Theory. It can be shown that all nondominated solutions will also be preferred solutions for some setting of the preference vector g. However, some preferred solutions may not be nondominated. The preferability relation can also be shown to be transitive. Detailed proofs can be found in the thesis of Fonseca (1995, pp 154–9).

5.4 Concluding remarks

The number and diversity of the multiobjective approaches to evolutionary optimization proposed to date is a clear sign of the growing interest and recognition this area is receiving. In contrast, quantitatively characterizing their expected performance, even on specific examples, has remained difficult, mainly due to the lack of a unique solution to such problems and to the number of performance dimensions involved. Needless to say, this has considerably impaired the realization of extensive comparative studies.

In the light of recent results concerning the performance assessment and comparison of multiobjective optimizers such as, but not limited to, evolutionary algorithms (Fonseca and Fleming 1996), this situation may soon change. Until then, the choice of a multiobjective fitness evaluation approach should take into account how much preference information is available for a particular problem, and in what form, as well the ease of implementation and whether or not it is possible or desirable to interact with the algorithm as it runs.

References

Ben-Tal A 1980 Characterization of Pareto and lexicographic optimal solutions *Multiple Criteria Decision Making Theory and Application (Lecture Notes in Economics and Mathematical Systems 177)* pp 1–11

Breeden J L 1995 Optimizing stochastic and multiple fitness functions *Proc. 4th Annu. Conf. on Evolutionary Programming (San Diego, CA, 1995)* ed J R McDonnel, R G Reynolds and D B Fogel (Cambridge, MA: MIT Press)

Cieniawski S E 1993 *An Investigation of the Ability of Genetic Algorithms to Generate the Tradeoff Curve of a Multi-objective Groundwater Monitoring Problem* Master's Thesis, University of Illinois at Urbana-Champaign

Fleming P J and Pashkevich A 1985 Computer aided control system design using a multiobjective optimization approach *Proc. IEE Control'85 Conf. (Cambridge, 1985)* pp 174–9

Fonseca C M 1995 *Multiobjective Genetic Algorithms with Application to Control Engineering Systems* PhD Thesis, University of Sheffield

Fonseca C M and Fleming P J 1993 Genetic algorithms for multiobjective optimization: formulation, discussion and generalization *Genetic Algorithms: Proc. 5th Int. Conf. (Urbana-Champaign, IL, 1993)* ed S Forrest (San Mateo, CA: Morgan Kaufmann) pp 416–23

——1995 An overview of evolutionary algorithms in multiobjective optimization *Evolutionary Comput.* **3** 1–16

——1996 On the performance assessment and comparison of stochastic multiobjective optimizers *Parallel Problem Solving from Nature—PPSN IV* ed H-M Voigt, W Ebeling, I Rechenberg and H-P Schwefel (Berlin: Springer) pp 584–93

Fourman M P 1985 Compaction of symbolic layout using genetic algorithms *Genetic Algorithms and Their Applications: Proc. 1st Int. Conf. on Genetic Algorithms* ed J J Grefenstette (Hillsdale, NJ: Erlbaum) pp 141–53

Gembicki F W 1974 *Vector Optimization for Control with Performance and Parameter Sensitivity Indices* PhD Thesis, Case Western Reserve University

Goicoechea A, Hansen D R and Duckstein L 1982 *Multiobjective Decision Analysis with Engineering and Business Applications* (New York: Wiley) p 46

Goldberg D E 1989 *Genetic Algorithms in Search, Optimization and Machine Learning* (Reading, MA: Addison-Wesley)

Hajela P and Lin C-Y 1992 Genetic search strategies in multicriterion optimal design *Struct. Optimization* **4** 99–107

Horn J, Nafpliotis N and Goldberg D E 1994 A niched Pareto genetic algorithm for multiobjective optimization *Proc. 1st IEEE Conf. on Evolutionary Computation,*

IEEE World Congr. on Computational Intelligence (Orlando, FL, 1994) (Piscataway, NJ: IEEE) pp 82–7

Kursawe F 1991 A variant of evolution strategies for vector optimization *Parallel Problem Solving from Nature, 1st Workshop (Lecture Notes in Computer Science 496)* ed H-P Schwefel and R Männer (Berlin: Springer) pp 193–7

Richardson J T, Palmer M R, Liepins G and Hilliard M 1989 Some guidelines for genetic algorithms with penalty functions *Proc. 3rd Int. Conf. on Genetic Algorithms (Fairfax, VA, 1989)* ed J D Schaffer (San Mateo, CA: Morgan Kaufmann) pp 191–7

Ritzel B J, Eheart J W and Ranjithan S 1994 Using genetic algorithms to solve a multiple objective groundwater pollution containment problem *Water Resources Res.* **30** 1589–603

Schaffer J D 1985 Multiple objective optimization with vector evaluated genetic algorithms *Genetic Algorithms and Their Applications: Proc. 1st Int. Conf. on Genetic Algorithms* ed J J Grefenstette (Hillsdale, NJ: Erlbaum) pp 93–100

Srinivas N and Deb K 1994 Multiobjective optimization using nondominated sorting in genetic algorithms *Evolutionary Comput.* **2** 221–48

Wienke D, Lucasius C and Kateman G 1992 Multicriteria target vector optimization of analytical procedures using a genetic algorithm Part I. Theory, numerical simulations and application to atomic emission spectroscopy *Anal. Chim. Acta* **265** 211–25

Wilson P B and Macleod M D 1993 Low implementation cost IIR digital filter design using genetic algorithms *IEE/IEEE Workshop on Natural Algorithms in Signal Processing (Chelmsford)* pp 4/1–8

6

Introduction to constraint-handling techniques

Zbigniew Michalewicz

In general, constraints are an integral part of the formulation of any problem. Dhar and Ranganathan (1990) wrote:

> ...Virtually all decision making situations involve constraints. What distinguishes various types of problems is the form of these constraints. Depending on how the problem is visualized, they can arise as rules, data dependencies, algebraic expressions, or other forms.
>
> Constraint satisfaction problems (CSPs) have been studied extensively in the operations research (OR) and artificial intelligence (AI) literature. In OR formulations constraints are quantitative, and the solver (such as the Simplex algorithm) optimizes (maximizes or minimizes) the value of a specified objective function subject to the constraints. In contrast, AI research has focused on inference-based approaches with mostly symbolic constraints. The inference mechanisms employed include theorem provers, production rule interpreters, and various labeling procedures such as those used in truth maintenance systems.

For example, in continuous domains, the general nonlinear programming problem is to find x so as to

$$\text{optimize } f(x) \qquad x = (x_1, \ldots, x_n) \in \mathbb{R}^n$$

where $x \in \mathcal{F} \subseteq \mathcal{S}$. The set $\mathcal{S} \subseteq \mathbb{R}^n$ defines the search space and the set $\mathcal{F} \subseteq \mathcal{S}$ defines a *feasible* search space. The search space $\mathcal{S}$ is defined as an n-dimensional rectangle in $\mathbb{R}^n$ (domains of variables defined by their lower and upper bounds):

$$l(i) \leq x_i \leq u(i) \qquad 1 \leq i \leq n$$

whereas the feasible set $\mathcal{F}$ is defined by an intersection of $\mathcal{S}$ and a set of additional $m \geq 0$ constraints:

$$g_j(x) \leq 0 \quad \text{for } j = 1, \ldots, q \qquad h_j(x) = 0 \quad \text{for } j = q + 1, \ldots, m.$$

In discrete domains, most problems are constrained: for example, the knapsack problem, set covering problem, vehicle routing problem, and all types of scheduling and timetabling problem are constrained.

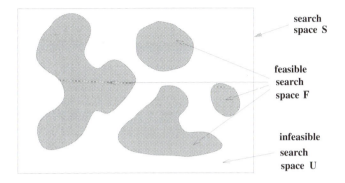

Figure 6.1. A search space and its feasible and infeasible parts.

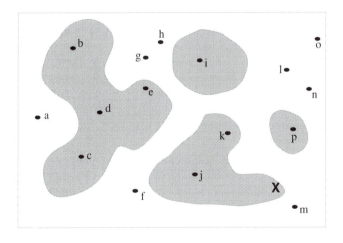

Figure 6.2. A population of 16 individuals, a–p.

In general, a search space $\mathcal{S}$ consists of two disjoint subsets of feasible and infeasible subspaces, $\mathcal{F}$ and $\mathcal{U}$, respectively (see figure 6.1). These subspaces need not be convex and they need not be connected (as, for example, in figure 6.1 where the feasible part $\mathcal{F}$ of the search space consists of four disjoined subsets). In solving optimization problems we search for a *feasible* optimum. During the search process we have to deal with various feasible and infeasible individuals; for example (see figure 6.2), at some stage of the evolution process, a population may contain some feasible (b, c, d, e, i, j, k, p) and infeasible individuals (a, f, g, h, l, m, n, o), while the optimum solution is marked by X.

The problem of how to deal with infeasible individuals is far from trivial. In general, we have to design two evaluation functions, eval_f and eval_u, for feasible and infeasible domains, respectively:

$$\text{eval}_f : \mathcal{F} \to \mathbb{R} \qquad \text{and} \qquad \text{eval}_u : \mathcal{U} \to \mathbb{R}.$$

There are many important questions to be addressed; these include:

- Should we choose to penalize infeasible individuals? In other words, should we extend the domain of function eval_f and assume that $\text{eval}_u(n) = \text{eval}_f(n) + \text{penalty}(n)$? (In the following discussion we continue referring to individuals just by plain letters of the alphabet, as displayed in figure 6.2; the reader should keep in mind that an individual, say, n, may represent a vector x in high-dimensional space). If so, how should such a penalty function penalty(n) be designed? In particular, should we consider infeasible individuals harmful and eliminate them from the population (Chapter 7)?
- Should we change the topology of the search space by using decoders, which interpret (transform) an individual into a feasible one (Chapter 8)?
- Should we 'repair' infeasible solutions by moving them into the closest point of the feasible space (e.g. the repaired version of m might be optimum X, figure 6.2)? In other words, should we assume that $\text{eval}_u(m) = \text{eval}_f(s)$, where s is a repaired version of m? If so, should we replace m by its repaired version s in the population or rather should we use a repair procedure for evaluation purposes only (Chapter 9)?
- Should we start with an initial population of feasible individuals and maintain the feasibility of offspring by using specialized operators (Chapter 10)?
- Should we use process solutions and constraints separately, or use cultural algorithms, or use co-evolutionary methods, that is, should we use some other, nonstandard constraint-handling technique (Chapter 11)?
- How should we locate feasible solutions (Chapter 12)?

The above questions are addressed in the following Chapters, thus providing a detailed discussion on various aspects of constraint-handling techniques.

Reference

Dhar V and Ranganathan N 1990 Integer programming vs. expert systems: an experimental comparison *Commun. ACM* **33** 323–36

7

Penalty functions

Alice E Smith and David W Coit

7.1 Introduction to penalty functions

Penalty functions have been a part of the literature on constrained optimization for decades. Two basic types of penalty function exist: exterior penalty functions, which penalize infeasible solutions, and interior penalty functions, which penalize feasible solutions. It is the former type of penalty function which is discussed throughout this section; however the area of interior penalty functions is of potential research interest in evolutionary computation. The main idea of interior penalty functions is that an optimal solution requires that a constraint be active (i.e. tight) so that this optimal solution lies on the boundary between feasibility and infeasibility. Knowing this, a penalty is applied to feasible solutions when the constraint is not active (such solutions are called *interior solutions*). For a single constraint, this approach is straightforward (although it has not been seen in the evolutionary computation literature); however, for the more common case of multiple constraints, the implementation of interior penalty functions is considerably more complex.

Three degrees of exterior penalty functions exist: (i) barrier methods in which no infeasible solution is considered, (ii) partial penalty functions in which a penalty is applied near the feasibility boundary, and (iii) global penalty functions that are applied throughout the infeasible region (Schwefel 1995, p 16). In the area of combinatorial optimization, the popular Lagrangian relaxation method (Avriel 1976, Fisher 1981, Reeves 1993) is a variation on the same theme: temporarily relax the problem's most difficult constraints, using a modified objective function to avoid straying too far from the feasible region. In general, a penalty function approach is as follows. Given an optimization problem, the following is the most general formulation of constraints:

$$\min f(x) \quad \text{such that } x \in A, x \in B \tag{7.1}$$

where x is a vector of decision variables, the constraints '$x \in A$' are relatively easy to satisfy, and the constraints '$x \in B$' are relatively difficult to satisfy; the problem can be reformulated as

$$\min f(x + p(d(x, B))) \quad \text{such that } x \in A \qquad (7.2)$$

where $d(x, B)$ is a metric function describing the distance of the solution vector x from the region B, and $p(\cdot)$ is a monotonically nondecreasing penalty function such that $p(0) = 0$. If the exterior penalty function, $p(\cdot)$, grows quickly enough outside of B, the optimal solution of (7.1) will also be optimal for (7.2). Furthermore, any optimal solution of (7.2) will (again, if $p(\cdot)$ grows quickly enough) provide an upper bound on the optimum for (7.1), and this bound will in general be tighter than that obtained by simply optimizing $f(x)$ over A.

In practice, the constraints '$x \in B$' are expressed as inequality and equality constraints in the form of

$$g_i(x) \le 0 \qquad \text{for } i = 1, \ldots, q$$
$$h_i(x) = 0 \qquad \text{for } i = q + 1, \ldots, m$$

where

$$q = \text{number of inequality constraints}$$
$$m - q = \text{number of equality constraints.}$$

Various families of functions $p(\cdot)$ and $d(\cdot)$ have been studied for evolutionary optimization to dualize constraints. Different possible distance metrics, $d(\cdot)$, include a count of the number of violated constraints, the Euclidean distance between x and B as suggested by Richardson et al (1989), a linear sum of the individual constraint violations or a sum of the individual constraint violations raised to an exponent, κ. Variations of these approaches have been attempted with different degrees of success. Some of the more notable examples are described in the following sections.

It can be difficult to find a penalty function that is an effective and efficient surrogate for the missing constraints. The effort required to tune the penalty function to a given problem instance or repeatedly calculate it during search may negate any gains in eventual solution quality. As noted by Siedlecki and Sklansky (1989), much of the difficulty arises because the optimal solution will frequently lie on the boundary of the feasible region. Many of the solutions most similar to the genotype of the optimum solution will be infeasible. Therefore, restricting the search to only feasible solutions or imposing very severe penalties makes it difficult to find the schemata that will drive the population toward the optimum as shown in the research of Smith and Tate (1993), Anderson and Ferris (1994), Coit et al (1996), and Michalewicz (1995). Conversely, if the penalty is not severe enough, then too large a region is searched and much of the search time will be used to explore regions far from the feasible region. Then, the search will tend to stall outside the feasible region. A good comparison of six penalty function strategies applied to continuous optimization problems is given by Michalewicz (1995). These strategies include both static and dynamic approaches, as discussed below, as well as some less generic approaches such as sequential constraint handling (Schoenauer and Xanthakis 1993) and forcing all infeasible solutions to be dominated by all feasible solutions in a given generation (Powell and Skolnick 1993).

7.2 Static penalty functions

A simple method to penalize infeasible solutions is to apply a constant penalty to those solutions that violate feasibility in any way. The penalized objective function would then be the unpenalized objective function plus a penalty (for a minimization problem). A variation is to construct this simple penalty function as a function of the number of constraints violated, where there are multiple constraints. The penalty function for a problem with m constraints would then be as below (for a minimization problem):

$$f_{\mathrm{p}}(\boldsymbol{x}) = f(\boldsymbol{x}) + \sum_{i=1}^{m} C_i \delta_i \qquad \text{where } \begin{cases} \delta_i = 1 & \text{if constraint } i \text{ is violated} \\ \delta_i = 0 & \text{if constraint } i \text{ is satisfied.} \end{cases}$$

$$(7.3)$$

$f_{\mathrm{p}}(\boldsymbol{x})$ is the penalized objective function, $f(\boldsymbol{x})$ is the unpenalized objective function, and C_i is a constant imposed for violation of constraint i. This penalty function is based only on the number of constraints violated, and is generally inferior to the second approach, based on some distance metric from the feasible region (Goldberg 1989, Richardson $et\ al$ 1989).

More common and more effective is to penalize according to distance to feasibility, or the 'cost to completion', as termed by Richardson $et\ al$ (1989). This was done crudely in the constant penalty functions of the preceding paragraph by assuming distance can be stated solely by number of constraints violated. A more sophisticated and more effective penalty includes a distance metric for each constraint, and adds a penalty that becomes more severe with distance from feasibility. Complicating this approach is the assumption that the distance metric chosen appropriately provides information concerning the nearness of the solution to feasibility, and the further implicit assumption that this nearness to feasibility is relevant in the same magnitude to the fitness of the solution. Distance metrics can be continuous (see, for example, Juliff 1993) or discrete (see, for example, Patton $et\ al$ 1995), and could be linear or nonlinear (see, for example, Le Riche $et\ al$ 1995).

A general formulation is as follows for a minimization problem:

$$f_{\mathrm{p}}(\boldsymbol{x}) = f(\boldsymbol{x}) + \sum_{i=1}^{m} C_i d_i^{\kappa} \qquad \text{where } d_i = \begin{cases} \delta_i g_i(\boldsymbol{x}) & \text{for } i = 1, \ldots, q \\ |h_i(\boldsymbol{x})| & \text{for } i = q+1, \ldots, m. \end{cases}$$

$$(7.4)$$

d_i is the distance metric of constraint i applied to solution x and κ is a user-defined exponent, with values of κ of one or two often used. Constraints 1–q are inequality constraints, so the penalty will only be activated when the constraint is violated (as shown by the δ function above), while constraints $(q+1)$–m are equality constraints which will activate the penalty if there is any distance between the solution value and the constraint value (as shown in the absolute distance above). In equation (7.4) above, defining C_i is more difficult. The advice from Richardson *et al* (1989) is to base C_i on the expected or maximum cost to repair the solution (i.e. alter the solution so it is feasible). For most problems, however, it is not possible to determine C_i using this rationale. Instead, it must be estimated based on the relative scaling of the distance metrics of multiple constraints, the difficulty of satisfying a constraint, and the seriousness of a constraint violation, or be determined experimentally.

Many researchers in evolutionary computation have explored variations of distance-based static penalty functions (e.g. Baeck and Khuri 1994, Goldberg 1989, Huang *et al* 1994, Olsen 1994, Richardson *et al* 1989). One example (Thangiah 1995) uses a linear combination of three constant distance-based penalties for the three constraints of the vehicle routing with time windows problem. Another novel example is from Le Riche *et al* (1995) where two separate distance-based penalty functions are used for each constraint in two genetic algorithm segregated subpopulations. This 'double penalty' somewhat improved robustness to penalty function parameters since the feasible optimum is approached with both a severe and a lenient penalty. Homaifar *et al* (1994) developed a unique static penalty function with multiple violation levels established for each constraint. Each interval is defined by the relative degree of constraint violation. For each interval l, a unique constant, C_{il}, is then used as a penalty function coefficient. This approach has the considerable disadvantage of requiring iterative tuning through experimentation of a large number of parameters.

7.3 Dynamic penalty functions

The primary deficiency with static penalty functions is the inability of the user to determine criteria for the C_i coefficients. Also, there are conflicting objectives involved with allowing exploration of the infeasible region, yet still requiring that the final solution be feasible. A variation of distance-based penalty functions that alleviates many of these difficulties is to incorporate a dynamic aspect that (generally) increases the severity of the penalty for a given distance as the search progresses. This has the property of allowing highly infeasible solutions early in the search, while continually increasing the penalty imposed to eventually move the final solution to the feasible region. A general form of a distance-based penalty method incorporating a dynamic aspect based on length of search, t, is as follows for a minimization problem:

$$f_p(\boldsymbol{x}, t) = f(\boldsymbol{x}) + \sum_{i=1}^{m} s_i(t) d_i^{\kappa} \tag{7.5}$$

where $s_i(t)$ is a function monotonically nondecreasing in value with t. Metrics for t include number of generations or the number of solutions searched. Recent uses of this approach include the work of Joines and Houck (1994) for continuous function optimization and Olsen (1994) and Michalewicz and Attia (1994), which compare several penalty functions, all of which consider distance, but some also consider evolution time. A common objective of these dynamic penalty formulations is that they result in feasible solutions at the end of evolution. If $s_i(t)$ is too lenient, final infeasible solutions may result, and if $s_i(t)$ is too severe, the search may converge to nonoptimal feasible solutions. Therefore, these penalty functions typically require problem-specific tuning to perform well. One explicit example of $s_i(t)$ is as follows, from Joines and Houck (1994):

$$s_i(t) = (C_i t)^{\alpha}$$

where α is constant equal to one or two, as defined by Joines and Houck.

7.4 Adaptive penalty functions

While incorporating distance together with the length of the search into the penalty function has been generally effective, these penalties ignore any other aspects of the search. In this respect, they are not adaptive to the ongoing success (or lack thereof) of the search and cannot guide the search to particularly attractive regions or away from unattractive regions based on what has already been observed. A few authors have proposed making use of such search-specific information. Siedlecki and Sklansky (1989) discuss the possibility of adaptive penalty functions, but their method is restricted to binary-string encodings with a single constraint, and involves considerable computational overhead.

Bean and Hadj-Alouane (1992) and Hadj-Alouane and Bean (1992) propose penalty functions that are revised based on the feasibility or infeasibility of the best, penalized solution during recent generations. Their penalty function allows either an increase or a decrease of the imposed penalty during evolution as shown below, and was demonstrated on multiple-choice integer programming problems with one constraint. This involves the selection of two constants, β_1 and β_2 ($\beta_1 > \beta_2 > 1$), to adaptively update the penalty function multiplier, and the evaluation of the feasibility of the best solution over successive intervals of N_f generations. As the search progresses, the penalty function multiplier is updated every N_f generations based on whether or not the best solution was feasible during that interval. Specifically, the penalty function is as follows;

$$f_p(\boldsymbol{x}, k) = f(\boldsymbol{x}) + \sum_{i=1}^{m} \lambda_k d_i^{\kappa}$$

$$\lambda_{k+1} = \begin{cases} \lambda_k \beta_1 & \text{if previous } N_f \text{ generations have} \\ & \text{only infeasible best solution} \\ \lambda_k / \beta_2 & \text{if previous } N_f \text{ generations have} \\ & \text{only feasible best solution} \\ \lambda_k & \text{otherwise.} \end{cases} \tag{7.6}$$

Smith and Tate (1993) and Tate and Smith (1995) used both search length and constraint severity feedback in their penalty function, which was enhanced by the work of Coit *et al* (1996). This penalty function involves the estimation of a near-feasible threshold (NFT) for each constraint. Conceptually, the NFT is the threshold distance from the feasible region at which the user would consider the search as 'getting warm'. The penalty function encourages the evolutionary algorithm to explore within the feasible region and the NFT neighborhood of the feasible region, and discourage search beyond that threshold. This formulation is given below:

$$f_p(\boldsymbol{x}, t) = f(\boldsymbol{x}) + (F_{\text{feas}}(t) - F_{\text{all}}(t)) \sum_{i=1}^{m} \left(\frac{d_i}{\text{NFT}_i} \right)^{\kappa} \tag{7.7}$$

where $F_{\text{all}}(t)$ denotes the unpenalized value of the best solution yet found, and $F_{\text{feas}}(t)$ denotes the value of the best feasible solution yet found. The $F_{\text{all}}(t)$ and $F_{\text{feas}}(t)$ terms serve several purposes. First, they provide adaptive scaling of the penalty based on the results of the search. Second, they combine with the NFT_i term to provide a search-specific and constraint-specific penalty.

The general form of NFT_i is

$$\text{NFT}_i = \frac{\text{NFT}_{0i}}{1 + \Lambda_i} \tag{7.8}$$

where NFT_{0i} is an upper bound for NFT_i. Λ_i is a dynamic search parameter used to adjust NFT_i based on the search history. In the simplest case, Λ_i can be set to zero and a static NFT_i results. Λ_i can also be defined as a function of the search, for example, a function of the generation number (t), i.e. $\Lambda_i = f(t) = \lambda_i t$. A positive value of λ_i results in a monotonically decreasing NFT_i (and, thus, a larger penalty) and a larger λ_i more quickly decreases NFT_i as the search progresses, incorporating both adaptive and dynamic elements.

If NFT_i is intuitively ill defined, it can be set at a large value initially with a positive constant λ_i used to iteratively guide the search to the feasible region. This dynamic NFT_i circumvents the need to perform experimentation to determine appropriate penalty function parameter values. However, if problem-specific information is at hand, a more efficient search can take place by *a priori* defining a tighter region or even static values of NFT_i.

7.5 Future directions in penalty functions

Two areas requiring further research are the development of completely adaptive penalty functions that require no user-specified constants and the development of improved adaptive operators to exploit characteristics of the search as they are found. The notion of adaptiveness is to leverage the information gained during evolution to improve both the effectiveness and the efficiency of the penalty function used. Another area of interest is to explore the assumption that multiple constraints can be linearly combined to yield an appropriate penalty function. This implicit assumption of all penalty functions used in the literature assumes that constraint violations incur independent penalties and therefore there is no interaction between constraints. Intuitively, this seems to be a possibly erroneous assumption, and one could make a case for a penalty that increases more than linearly with the number of constraints violated.

References

Anderson E J and Ferris M C 1994 Genetic algorithms for combinatorial optimization: the assembly line balancing problem *ORSA J. Comput.* **6** 161–73

Avriel M 1976 *Nonlinear Programming: Analysis and Methods* (Englewood Cliffs, NJ: Prentice-Hall)

Baeck T and Khuri S 1994 An evolutionary heuristic for the maximum independent set problem *Proc. 1st IEEE Conf. on Evolutionary Compution (Orlando, FL, June 1994)* (Piscataway, NJ: IEEE) pp 531–5

Bean J C and Hadj-Alouane A B 1992 *A Dual Genetic Algorithm for Bounded Integer Programs* University of Michigan Technical Report 92-53; Revue francaise d'automatique, d'informatique et de recherche operationnelle: Recherche operationnelle, at press (in French)

Coit D W, Smith A E and Tate D M 1996 Adaptive penalty methods for genetic optimization of constrained combinatorial problems *INFORMS J. Comput.* **8** 173–82

Fisher M L 1981 The Lagrangian relaxation method for solving integer programming problems *Management Sci.* **27** 1–18

Goldberg D E 1989 *Genetic Algorithms in Search Optimization and Machine Learning* (Reading, MA: Addison-Wesley)

Hadj-Alouane A B and Bean J C 1992 *A Genetic Algorithm for the Multiple-Choice Integer Program* University of Michigan Technical Report 92-50; *Operations Res.* at press

Homaifar A, Lai S H-Y and Qi Z 1994 Constrained optimization via genetic algorithms *Simulation* **62** 242–54

Huang W-C, Kao C-Y and Horng J-T 1994 A genetic algorithm approach for set covering problem *Proc. 1st IEEE Conf. on Evolutionary Computation (Orlando, FL, June 1994)* (Piscataway, NJ: IEEE) pp 569–73

Joines J A and Houck C R 1994 On the use of non-stationary penalty functions to solve nonlinear constrained optimization problems with GA's *Proc. 1st IEEE Conf. on Evolutionary Computation (Orlando, FL, 1994)* (Piscataway, NJ: IEEE) pp 579–84

Juliff K 1993 A multi-chromosome genetic algorithm for pallet loading *Proc. 5th Int. Conf. on Genetic Algorithms (Urbana-Champaign, IL, July 1993)* ed S Forrest (San Mateo, CA: Morgan Kaufmann) pp 467–73

Le Riche R G, Knopf-Lenoir C and Haftka R T 1995 A segregated genetic algorithm for constrained structural optimization *Proc. 6th Int. Conf. on Genetic Algorithms (Pittsburgh, PA, July 1995)* ed L J Eshelman (San Mateo, CA: Morgan Kaufmann) pp 558–65

Michalewicz Z 1995 Genetic algorithms numerical optimization and constraints *Proc. 6th Int. Conf. on Genetic Algorithms (Pittsburgh, PA, July 1995)* ed L J Eshelman (San Mateo, CA: Morgan Kaufmann) pp 151–8

Michalewicz Z and Attia N 1994 Evolutionary optimization of constrained problems *Proc. 3rd Ann. Conf. on Evolutionary Programming (San Diego, CA, February 1994)* ed A V Sebald and L J Fogel (Singapore: World Scientific) pp 98–108

Olsen A L 1994 Penalty functions and the knapsack problem *Proc. 1st IEEE Conf. on Evolutionary Computation (Orlando, FL, June 1994)* (Piscataway, NJ: IEEE) pp 554–8

Patton A L, Punch W F III and Goodman E D 1995 A standard GA approach to native protein conformation prediction *Proc. 6th Int. Conf. on Genetic Algorithms (Pittsburgh, PA, July 1995)* ed L J Eshelman (San Mateo, CA: Morgan Kaufmann) pp 574–81

Powell D and Skolnick M M 1993 Using genetic algorithms in engineering design optimization with non-linear constraints *Proc. 5th Int. Conf. on Genetic Algorithms (Urbana-Champaign, IL, July 1993)* ed S Forrest (San Mateo, CA: Morgan Kaufmann) pp 424–30

Reeves C R 1993 *Modern Heuristic Techniques for Combinatorial Problems* (New York: Wiley)

Richardson J T, Palmer M R, Liepins G and Hilliard M 1989 Some guidelines for genetic algorithms with penalty functions *Proc. 3rd Int. Conf. on Genetic Algorithms (Fairfax, VA, June 1989)* ed J D Schaffer (San Mateo, CA: Morgan Kaufmann) pp 191–7

Schoenauer M and Xanthakis S 1993 Constrained GA optimization *Proc. 5th Int. Conf. on Genetic Algorithms (Urbana-Champaign, IL, July 1993)* ed S Forrest (San Mateo, CA: Morgan Kaufmann) pp 573–80

Schwefel H-P 1995 *Evolution and Optimum Seeking* (New York: Wiley)

Siedlecki W and Sklansky J 1989 Constrained genetic optimization via dynamic reward–penalty balancing and its use in pattern recognition *Proc. 3rd Int. Conf. on Genetic Algorithms (Fairfax, VA, June 1989)* ed J D Schaffer (San Mateo, CA: Morgan Kaufmann) pp 141–50

Smith A E and Tate D M 1993 Genetic optimization using a penalty function *Proc. 5th Int. Conf. on Genetic Algorithms (Urbana-Champaign, IL, July 1993)* ed S Forrest (San Mateo, CA: Morgan Kaufmann) pp 499–505

Tate D M and Smith A E 1995 Unequal area facility layout using genetic search *IIE Trans.* **27** 465–72

Thangiah S R 1995 An adaptive clustering method using a geometric shape for vehicle routing problems with time windows *Proc. 6th Int. Conf. on Genetic Algorithms (Pittsburgh, PA, July 1995)* ed L J Eshelman (San Mateo, CA: Morgan Kaufmann) pp 536–43

8

Decoders

Zbigniew Michalewicz

8.1 Introduction

Decoders offer an interesting option for all practitioners of evolutionary techniques. In these techniques a chromosome 'gives instructions' to a decoder or 'is interpreted' by a decoder on how to build a feasible solution. For example, a sequence of items for the knapsack problem can be interpreted as 'take an item if possible'—such interpretation would lead always to feasible solutions. Let us consider the following scenario: we try to solve the 0–1 knapsack problem (see Chapter 9 for a brief description of this problem) with n items; the profit and weight of the ith item are p_i and w_i, respectively. We can sort all items in decreasing order of p_i/w_i values and interpret the binary string

$$(110011000100111010100101011101010101\ldots 0010)$$

in the following way. Take the first item from the list (i.e. the item with the largest ratio of profit per weight) if the item fits in the knapsack. Continue with the second, fifth, sixth, tenth, and so on, items from the sorted list (i.e. continue with items with corresponding 1's in the binary string), until the knapsack is full or there are no more items available (note that the binary string of all 1's corresponds to a greedy solution). Any sequence of bits would translate into a feasible solution; any feasible solution may have many possible codes (which may differ in the rightmost string part). We can apply classical binary operators (crossover and mutation): any offspring is clearly feasible.

8.2 The traveling salesman problem

A similar approach has been tried for solving the traveling salesman problem (Grefenstette *et al* 1985). For example, a chromosome may represent a tour as a list of n cities; the ith element of the list is a number in the range from 1 to $n-i+1$. The idea behind such decoders is as follows. There is some ordered list of cities C, which serves as a reference point for lists in ordinal representations (like a sorted sequence of items for the knapsack problem discussed earlier). Assume, for example, that such an ordered list (a reference point) is simply

$$C = (1\ 2\ 3\ 4\ 5\ 6\ 7\ 8\ 9).$$

A tour

$$1\text{--}2\text{--}4\text{--}3\text{--}8\text{--}5\text{--}9\text{--}6\text{--}7$$

is then represented as a list l of references,

$$l = (1\ 1\ 2\ 1\ 4\ 1\ 3\ 1\ 1)$$

and should be interpreted as follows: the first number on the list l is 1, so take the first city from the list C as the first city of the tour (city number 1), and remove it from C. At this stage the partial tour is (1). The next number on the list l is also 1, so take the first city from the current list C as the next city of the tour (city number 2), and remove it from C. At this stage the partial tour is (1, 2), and so on. The main advantage of the ordinal representation is that the classical crossover works: any two tours in the ordinal representation, cut after some position and crossed together, would produce two offspring, each of them being a legal tour. For example, the two parents

$$p_1 = (1\ 1\ 2\ 1\ |\ 4\ 1\ 3\ 1\ 1)$$

and

$$p_2 = (5\ 1\ 5\ 5\ |\ 5\ 3\ 3\ 2\ 1)$$

which correspond to the tours

$$1\ \text{--}2\text{--}4\text{--}3\text{--}8\text{--}5\text{--}9\text{--}6\text{--}7$$

and

$$5\text{--}1\text{--}7\text{--}8\text{--}9\text{--}4\text{--}6\text{--}3\text{--}2$$

with the crossover point marked by |, would produce the following offspring:

$$o_1 = (1\ 1\ 2\ 1\ 5\ 3\ 3\ 2\ 1)$$

and

$$o_2 = (5\ 1\ 5\ 5\ 4\ 1\ 3\ 1\ 1).$$

These offspring correspond to

$$1\text{--}2\text{--}4\text{--}3\text{--}9\text{--}7\text{--}8\text{--}6\text{--}5$$

and

$$5\text{--}1\text{--}7\text{--}8\text{--}6\text{--}2\text{--}9\text{--}3\text{--}4.$$

There are many other examples of how decoders have been used for a particular application. These include work on scheduling problems (see, for example, Bagchi *et al* 1991 and Syswerda 1991), pallet loading (Juliff 1993) and partitioning (Jones and Beltramo 1991).

8.3 Formal description

More formally, a decoder is a mapping T from a representation space (e.g. a space of binary strings, vectors of integer numbers and the like) into a feasible part of the solution space $\mathcal{F}$—viewing decoders from this perspective, evolutionary computation technique with a decoder is identical to so-called

Figure 8.1. Transformation T between solutions in (a) original and (b) decoder's space.

morphogenic evolutionary techniques (Angeline 1995), which include mappings (i.e. development functions) between representations that evolve (i.e. evolved representations) and representations that constitute the input for the evaluation function (i.e. evaluated representation). A graphical example of such a mapping is given in figure 8.1, where the mapping T transforms a point d in the representation space (part (b)) into a feasible solution s (part (a)).

However, it is important that several conditions are satisfied (Palmer and Kershenbaum 1994):

- for each solution $s \in \mathcal{F}$ there is a solution d from the representation space
- each solution d from the representation space corresponds to a feasible solution $s \in \mathcal{F}$
- all solutions in $\mathcal{F}$ should be represented by the same number of solutions (codings) d.

Additionally, it is reasonable to request that:

- the transformation T is computationally fast
- it has a locality feature in the sense that small changes in a solution from the representation space result in small changes in the (feasible) solution itself.

Also, as stated by Davis (1987),

> If one builds a 'decoder' into the evaluation procedure that intelligently avoids building an illegal individual from the chromosome, the result is frequently computation-intensive to run. Further, not all constraints can be easily implemented in this way.

For many years decoders have been used for discrete optimization problems only. Only recently an approach for solving constrained numerical optimization problems based on a mapping between n-dimensional cube and a feasible search space was described (Kozieland Michalewicz 1998, 1999). A mapping φ was developed, which transforms the n-dimensional cube $[-1, 1]^n$ into the feasible region $\mathcal{F}$ of the problem. Note that $\mathcal{F}$ need not be convex; it might be concave or even can consist of disjoint (non-convex) regions. The feasible region is defined by an arbitrary set of inequalities:

$$g_i(x) \leq 0 \qquad \text{for } i = 1, \ldots, m.$$

The mapping proposed by Kozieland Michalewicz (1999) was based on the following concepts. First, an additional one-to-one mapping g between the cube $[-1, 1]^n$ and the search space S was defined (note that the search space S is defined as a Cartesian product of domains of all problem variables; the domain of variable x_i is $[l(i), u(i)]$ for $i = 1, \ldots, n$). Then the mapping $g : [-1, 1]^n \rightarrow S$ can be defined as

$$g(y) = x \qquad \text{where } x_i = y_i \frac{u(i) - l(i)}{2} + \frac{u(i) + l(i)}{2} \quad \text{for } i = 1, \ldots, n.$$

Indeed, for $y_i = -1$ the corresponding $x_i = l(i)$, and for $y_i = 1$, $x_i = u(i)$.

Now, a line segment L between any reference point $r_0 \in \mathcal{F}$ and a point s at the boundary of the search space S is defined as

$$L(r_0, s) = r_0 + t(s - r_0) \qquad \text{for } 0 \leq t \leq 1.$$

Clearly, if the feasible search space $\mathcal{F}$ is convex, then the above line segment intersects the boundary of $\mathcal{F}$ at precisely one point, for some $t_0 \in [0, 1]$. Consequently, for convex feasible search spaces $\mathcal{F}$, it is possible to establish a one-to-one mapping $\varphi : [-1, 1]^n \rightarrow \mathcal{F}$ as follows:

$$\varphi(y) = \begin{cases} r_0 + y_{\max} t_0 (g(y/y_{\max}) - r_0) & \text{if } y \neq 0 \\ r_0 & \text{if } y = 0 \end{cases}$$

where $r_0 \in \mathcal{F}$ is a reference point, and $y_{\max} = \max_{i=1}^{n} |y_i|$. Figure 8.2 illustrates the transformation φ.

The mapping is more complex for nonconvex feasible search spaces $\mathcal{F}$. Let us consider an arbitrary point $y \in [-1, 1]^n$ and a reference point $r_0 \in \mathcal{F}$. A line segment L between the reference point r_0 and the point $s = g(y/y_{\max})$ at the boundary of the search space S is defined as before:

$$L(r_0, s) = r_0 + t(s - r_0) \qquad \text{for } 0 \leq t \leq 1$$

however, it may intersect the boundary of $\mathcal{F}$ at many points. In other words, instead of a single interval of feasibility $[0, t_0]$ for convex search spaces, we may have several intervals of feasibility:

$$[t_1, t_2], \ldots, [t_{2k-1}, t_{2k}].$$

Assume there are altogether k subintervals of feasibility for a such line segment and the t_i mark their limits. Clearly, $t_1 = 0$, $t_i < t_{i+1}$ for $i = 1, \ldots, 2k - 1$, and $t_{2k} \leq 1$ (see figure 8.3).

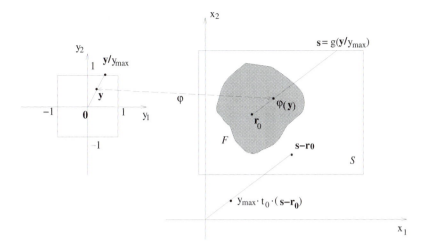

Figure 8.2. A mapping φ from the cube $[-1, 1]^n$ into the convex space $\mathcal{F}$ (two-dimensional case), with particular steps of the transformation.

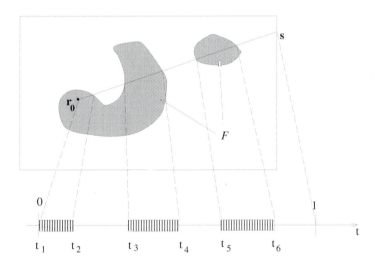

Figure 8.3. A line segment in a nonconvex space $\mathcal{F}$ and corresponding subintervals (two-dimensional case).

Thus, it is necessary to introduce an additional mapping γ, which transforms interval $[0, 1]$ into the sum of intervals $[t_{2i-1}, t_{2i}]$. However, we define such a mapping γ rather between $(0, 1]$ and the sum of intervals $(t_{2i-1}, t_{2i}]$:

$$\gamma : (0, 1] \to \bigcup_{i=1}^{k} (t_{2i-1}, t_{2i}].$$

Note that, due to this change, the left boundary point (from each interval $1 \leq i \leq k$) is lost. This is not a serious problem, since we can approach the lost points with arbitrary precision. On the other hand, the benefits are clear: it is possible to 'glue together' intervals which are open at one end and closed at another; additionally, such a mapping is one-to-one. There are many possibilities for defining such a mapping; we have used the following. First, let us define a reverse mapping δ:

$$\delta : \bigcup_{i=1}^{k} (t_{2i-1}, t_{2i}] \rightarrow (0, 1]$$

as follows:

$$\delta(t) = \left(t - t_{2i-1} + \sum_{j=1}^{i-1} d_j \right) / d$$

where $d_j = t_{2j} - t_{2j-1}$, $d = \sum_{j=1}^{k} d_j$, and $t_{2i-1} < t \leq t_{2i}$. Clearly, the mapping γ is the reverse of δ:

$$\gamma(a) = t_{2j-1} + d_j \frac{a - \delta(t_{2j-1})}{\delta(t_{2j}) - \delta(t_{2j-1})}$$

where j is the smallest index such that $a \leq \delta(t_{2j})$.

Now, the general mapping φ, a transformation of the constrained optimization problem to the unconstrained one for every feasible set $\mathcal{F}$, can be defined; it is given by the following formula:

$$\varphi(y) = \begin{cases} r_0 + t_0(g(y/y_{\max}) - r_0) & \text{if } y \neq 0 \\ r_0 & \text{if } y = 0 \end{cases}$$

where $r_0 \in \mathcal{F}$ is a reference point, $y_{\max} = \max_{i=1}^{n} |y_i|$, and $t_0 = \gamma(|y_{\max}|)$.

Finally, it is necessary to consider a method of finding such points of intersections t_i (see figure 8.3). This is relatively easy for convex sets, since there was only one point of intersection. In the implementation of this system (Kozieland Michalewicz 1999) the authors used the following approach. They considered any boundary point s of $\mathcal{S}$ and the line segment L determined by this point and a reference point $r_0 \in \mathcal{F}$. There are m constraints $g_i(x) \leq 0$ and each of them can be represented as a function β_i of one independent variable t (for fixed reference point $r_0 \in \mathcal{F}$ and the boundary point s of $\mathcal{S}$):

$$\beta_i(t) = g_i(L(r_0, s)) = g_i(r_0 + t(s - r_0)) \qquad \text{for } 0 \leq t \leq 1 \text{ and } i = 1, \dots, m.$$

As the feasible region need not be convex, it may have more than one point of intersection of the segment L with the boundaries of the set $\mathcal{F}$. Therefore, the interval $[0, 1]$ is partitioned into v subintervals $[v_{j-1}, v_j]$, where $v_j - v_{j-1} = 1/v$ $(1 \leq j \leq v)$, so that equations $\beta_i(t) = 0$ have at most one solution in every subinterval. In that case the points of intersection can be determined by a binary search. Once the intersection points between a line segment L and all constraints $g_i(x) \leq 0$ are known, it is quite easy to determine intersection points between this line segment L and the boundary of the feasible set $\mathcal{F}$. For the experimental results the reader is referred to the article by Kozieland Michalewicz (1999).

In (Kozieland Michalewicz, 1998) the authors investigated the role of the reference point r_0, which can 'follow' the best solution found so far (the method of iterative solution improvement). In that way, the reference point can adapt itself to the current state of the search. This approach can be combined also with some other improvements (e.g. a non-uniform distribution of values of t can concentrate the search around the reference point which follows the best solution).

References

Angeline P J 1995 Morphogenic evolutionary computation: introduction, issues, and examples *Proc. 4th Ann. Conf. on Evolutionary Programming (San Diego, CA, March 1995)* ed J R McDonnell, R G Reynolds and D B Fogel (Cambridge, MA: MIT Press) pp 387–401

Bagchi S, Uckun S, Miyabe Y and Kawamura K 1991 Exploring problem-specific recombination operators for job shop scheduling *Proc. 4th Int. Conf. on Genetic Algorithms (San Diego, CA, July 1991)* ed R Belew and L Booker (San Mateo, CA: Morgan Kaufmann) pp 10–7

Davis L (ed) 1987 *Genetic Algorithms and Simulated Annealing* (San Mateo, CA: Morgan Kaufmann)

Grefenstette J J, Gopal R, Rosmaita B and Van Gucht D 1985 Genetic algorithm for the TSP *Proc. 1st Int. Conf. on Genetic Algorithms (Pittsburgh, PA, July 1985)* ed J J Grefenstette (San Mateo, CA: Morgan Kaufmann) pp 160–8

Jones D R and Beltranto M A 1991 Solving partitioning problems with genetic algorithms *Proc. 4th Int. Conf. on Genetic Algorithms (San Diego, CA, July 1991)* ed R Belew and L Booker (San Mateo, CA: Morgan Kaufmann) pp 442–9

Juliff K 1993 A multi-chromosome genetic algorithm for pallet loading *Proc. 5th Int. Conf. on Genetic Algorithms (Urbana-Champaign, IL, July 1993)* ed S Forrest (San Mateo, CA: Morgan Kaufmann) pp 467–73

KozielS and Michalewicz Z 1998 A decoder-based evolutionary algorithm for constrained parameter optimization problems. *Proc. 5th Conf. on Parallel Problem Solving from Nature* (Lecture Notes in Computer Science 1498) ed A E Eiben, T Bäck, M Schoenauer and H-P Schwefel (Berlin: Springer) pp 231–240

——1999 Evolutionary algorithms, homomorphous mappings, and constrained parameter optimization *Evolutionary Comput.* **7** 19–44

Palmer C C and Kershenbaum A 1994 Representing trees in genetic algorithms *Proc. IEEE Int. Conf. on Evolutionary Computation (Orlando, FL, June 1994)* pp 379–84

Syswerda G 1991 Schedule optimization using genetic algorithms *Handbook of Genetic Algorithms* ed L Davis (New York: Van Nostrand Reinhold) pp 332–49

9

Repair algorithms

Zbigniew Michalewicz

9.1 Introduction

Repair algorithms enjoy a particular popularity in the evolutionary computation community: for many combinatorial optimization problems (e.g. the traveling salesman problem, the knapsack problem, and the set covering problem) it is relatively easy to *repair* an infeasible individual. Such a repaired version can be used either for evaluation only; that is,

$$\text{eval}_u(y) = \text{eval}_f(x)$$

where x is a repaired (i.e. feasible) version of y, or it can also replace (with some probability) the original individual in the population. Note that the repaired version of solution m (figure 6.2) might be the optimum X.

The process of repairing infeasible individuals is related to combination of learning and evolution (the so-called *Baldwin effect*, Whitley *et al* 1994). Learning (as local search in general, and local search for the closest feasible solution, in particular) and evolution interact with each other: the evaluation of the improvement (again, improvement in the sense of finding a repaired, feasible solution) is transferred to the individual. In this way a local search is analogous to learning that occurs during one generation of a particular string. Note that the repair process is used *only* for evaluation of an individual; the repaired version of the individual does not replace the original one.

The weakness of these methods is in their problem dependence. For each particular problem a specific repair algorithm should be designed. Moreover, there are no standard heuristics on design of such algorithms; usually it is possible to use a greedy repair or random repair or incorporate any other heuristic which would guide the repair process. Also, for some problems the process of repairing infeasible individuals may be as complex as solving the original problem. This is the case for the nonlinear transportation problem (see Michalewicz 1993), most scheduling and timetable problems, and many others.

The question of replacing repaired individuals is related to so-called Lamarckian evolution (Whitley *et al* 1994), which assumes that an individual improves during its lifetime and that the resulting improvements are coded back into the chromosome. As stated by Whitley *et al* (1994),

> Our analytical and empirical results indicate that Lamarckian strategies are often an extremely fast form of search. However, functions exist where both the simple genetic algorithm without learning and the Lamarckian strategy used ... converge to local optima while the simple genetic algorithm exploiting the Baldwin effect converges to a global optimum.

This is why it is necessary to use the replacement strategy very carefully.

Recently Orvosh and Davis (1993) reported a so-called 5% rule: this heuristic rule states that in many combinatorial optimization problems an evolutionary computation technique with a repair algorithm provides the best results when 5% of repaired individuals replace their infeasible originals. However, many recent experiments (see e.g. Michalewicz 1996) have indicated that for many combinatorial optimization problems this rule does not apply. Either a different percentage gives better results, or there is no significant difference in the performance of the algorithm for various probabilities of replacement.

It seems that the 'optimal' probability of replacement is problem dependent and it may change over the evolution process as well. Further research is required to compare different heuristics for setting this parameter, which is of great importance for all repair-based methods.

We shall illustrate the above points on two examples (taken from discrete and continuous domains): the 0–1 knapsack problem and the nonlinear programming problem.

The 0–1 knapsack problem can be formulated as follows: for a given set of weights w_i, profits p_i, and capacity C, find a binary vector $x = \langle x_1, \ldots, x_n \rangle$, such that

$$\sum_{i=1}^{n} x_i w_i \leq C$$

and for which

$$\mathcal{P}(x) = \sum_{i=1}^{n} x_i p_i$$

is maximum.

A binary string of the length n represents a solution x to the problem: the ith item is selected for the knapsack iff $x[i] = 1$. The evaluation of each string is determined on the basis of its feasibility; the evaluation measure eval_f for a feasible string x is

$$\text{eval}_f(x) = \sum_{i=1}^{n} x_i p_i$$

whereas the evaluation measure eval$_u$ for a infeasible string x is

$$\text{eval}_u(x) = \sum_{i=1}^{n} x_i' p_i$$

where vector x' is a repaired version of the original vector x.

The procedure for converting infeasible x into feasible x' is straightforward:

Input: x

Output: x', the repaired version of x

knapsack-overfilled ← false;
$x' \leftarrow x$
if $\sum_{i=1}^{n} x_i' w_i > C$
then
        knapsack-overfilled ← true
fi
while (knapsack-overfilled) **do**
        $i \leftarrow$ **select** an item from the knapsack
        remove the selected item from the knapsack:
                $x_i' \leftarrow 0;$
        if $\sum_{i=1}^{n} x_i' w_i \leq C$
        then
                knapsack-overfilled ← false
        fi
od

9.2   First example

There are still several possible repair methods which follow the outline of this repair procedure; they may differ in selection procedure **select**, which chooses an item for removal from the knapsack. For example, the procedure **select** (i) may select a random element from the knapsack, (ii) may select the first available element from the left (right) of the list, (iii) may sort all items in the knapsack in decreasing order of their profit to weight ratios and always choose the last item (from the list of available items) for deletion (i.e. greedy repair), or (iv) may sort all items in the knapsack in decreasing order of their profit to weight ratios and choose an item (from the list of available items) for deletion with respect to some probability distribution (items with a larger ratio would have smaller probability of selection). Other repair methods are also possible.

In general, there are two categories of repair methods. Some of them (such as (i) and (iv) in the previous paragraph) contain an element of randomness; consequently, it is possible that two identical solutions have different evaluation measures. On the other hand, other repair methods (such as (ii) or (iii) in the previous paragraph) are deterministic.

From experiments reported by Michalewicz (1996) it seems that deterministic (greedy) repair gives much better results than random repairs. Additionally, the experiments did not confirm the 5% replacement rule: either a different percentage gave better results, or there was no significant difference in the performance of the algorithm for various probabilities of replacement. In most cases, the higher the replacement ratio, the better the result (as a rule of a thumb, experiments with the 0–1 knapsack problem suggest a replacement ratio of 1.0!).

9.3 Second example

The second example of a repair process in evolutionary techniques is taken from a continuous domain: it is the nonlinear programming problem. The problem is formulated as follows: find x so as to

$$\text{optimize } f(x) \qquad x = (x_1, \ldots, x_n) \in \mathbb{R}^n$$

subject to

$$g_j(x) \leq 0 \quad \text{for } j = 1, \ldots, q \qquad \text{and} \qquad h_j(x) = 0 \quad \text{for } j = q + 1, \ldots, m.$$

Michalewicz and Nazhiyath (1995) reported on implementation of a new system, Genocop III. This maintains two separate populations, where a development in one population influences evaluations of individuals in the other population. The first population P_s consists of so-called search points. Search points need not be feasible; the variables just stay within specified limits, that is, they satisfy domain constraints. (In Genocop III it is also possible to define linear constraints as a separate set of constraints, which are handled by specialized operators.) The second population P_r consists of so-called reference points; these points are fully feasible, that is, they satisfy *all* constraints (if the system cannot find any reference point, the user is prompted for it).

Reference points r from P_r, being feasible, are evaluated directly by the objective function (i.e. $\text{eval}_f(r) = f(r)$). On the other hand, search points from P_s are repaired for evaluation and the repair process for $s \in P_s$ works as follows. If s is feasible, then $\text{eval}_f(s) = f(s)$. Otherwise (i.e. $s \notin \mathcal{F}$), the system selects one of the reference points, for example, r from P_r, and creates a sequence of points z_i from a segment between s and r: $z_i = a_i s + (1 - a_i)r$. This can be done either (i) in a random way by generating random numbers a_i from the range $\langle 0, 1 \rangle$, or (ii) in a deterministic way by setting $a_i = 1/2, 1/4, 1/8, \ldots$ until a feasible point is found. Figure 9.1 illustrates the point.

The system has a few additional parameters. As explained in the previous paragraph, two repair methods are available: random or deterministic. Also, it is possible to specify the way a reference point is selected for a repair process. This

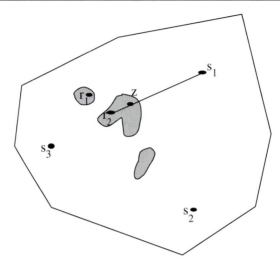

Figure 9.1. The repair process. A solution (search point) s_1 is repaired (point z) with respect to the reference solution r_2. The feasible areas of the search space are shaded.

selection is random either with a uniform probability distribution (all reference points have equal chances for selection) or with a probability distribution of reference points that depends on their evaluations (a ranking method is used). Clearly, in different generations the same search point S can evaluate to different values due to the random nature of the repair process.

Additionally, if $f(z)$ is better than $f(r)$, then the point z replaces r as a new reference point in the population of reference points P_r. Also, z replaces s in the population of search points P_s with some probability of replacement p_r.

It was interesting to check whether some $p\%$ rule (like the 5% rule of Orvosh and Davis 1993, reported for discrete domains) would emerge from experiments for numerical optimization problems. For this purpose one of the test problems for Genocop III was selected.

The problem (Keane 1994) is to maximize a function:

$$f(x) = \left| \frac{\sum_{i=1}^{n} \cos^4(x_i) - 2 \prod_{i=1}^{n} \cos^2(x_i)}{\left(\sum_{i=1}^{n} i x_i^2 \right)^{1/2}} \right|$$

where

$$\prod_{i=1}^{n} x_i > 0.75 \qquad \sum_{i=1}^{n} x_i < 7.5n \qquad 0 < x_i < 10 \quad \text{for } 1 \leq i \leq n.$$

Genocop III was run for the case of $n = 20$ for 10 000 generations with different values of replacement ratio. It was interesting to note the increase of the performance (in terms of the best solution found) of the system when the replacement ratio was increased gradually from 0.00 to 0.15%. For the ratio of 0.15% the best solution found was

$$x = (3.163\,113\,59, 3.131\,504\,30, 3.095\,158\,58, 3.060\,165\,88, 3.031\,035\,66,$$
$$2.991\,585\,49, 2.958\,025\,93, 2.922\,858\,95, 0.486\,843\,88, 0.477\,322\,79,$$
$$0.480\,444\,73, 0.487\,909\,11, 0.484\,504\,37, 0.448\,070\,32, 0.468\,777\,60,$$
$$0.456\,485\,06, 0.447\,626\,08, 0.449\,139\,86, 0.443\,908\,63, 0.451\,493\,32)$$

where $f(x) = 0.803\,510\,67$. However, further increases deteriorated the performance of the system; often the system converged to points $y \in \mathcal{F}$, where $0.75 \leq f(y) \leq 0.78$.

It is too early to claim a 15% replacement rule for continuous domains; however, $p_r = 0.15$ gave also the best results for other test cases.

9.4 Conclusion

Clearly, further research is necessary to investigate the relationship between optimization problems and repair techniques (these include repair methods as well as replacement rates).

References

Keane A 1994 *Genetic Algorithms Digest* **8** issue 16

Michalewicz Z 1993 A hierarchy of evolution programs: an experimental study *Evolutionary Comput.* **1** 51–76

——1996 *Genetic Algorithms + Data Structures = Evolution Programs* 3rd edn (New York: Springer)

Michalewicz Z and Nazhiyath G 1995 Genocop III: a co-evolutionary algorithm for numerical optimization problems with nonlinear constraints *Proc. 2nd IEEE Int. Conf. on Evolutionary Computation (Perth, 1995)* pp 647–51

Orvosh D and Davis L 1993 Shall we repair? Genetic algorithms, combinatorial optimization, and feasibility constraints *Proc. 5th Int. Conf. on Genetic Algorithms (Urbana-Champaign, IL, July 1993)* ed S Forrest (San Mateo, CA: Morgan Kaufmann) p 650

Whitley D, Gordon V S and Mathias K 1994 Lamarckian evolution, the Baldwin effect and function optimization *Proc. 3rd Conf. on Parallel Problem Solving from Nature (October 1994, Jerusalem)* ed Yu Davidor, H-P Schwefel and R Männer (Berlin: Springer) pp 6–15

10

Constraint-preserving operators

Zbigniew Michalewicz

10.1 Introduction

Many researchers have successfully experimented with specialized operators which preserve feasibility of individuals. These specialized operators incorporate problem-specific knowledge; the purpose of incorporating domain-based heuristics into operators (Surry *et al* 1995) is

> ... to build and use genetic operators that 'understand' the constraints, in the sense that they never produce infeasible solutions (ones that violate the constraints). ...The search is thus reformulated as an unconstrained optimization problem over the reduced space.

The main disadvantages connected with this approach are that (i) the problem-specific operators must be tailored for a particular application, and that (ii) it is very difficult to provide any formal analysis of such a system (however, important work towards understanding the way in which operators manipulate chromosomes is reported by Radcliffe (1991, 1994)). Nevertheless, there is overwhelming experimental evidence for the usefulness of this approach.

In this section we illustrate the case of problem-specific operators on three examples: the transportation problem, nonlinear optimization with linear constraints, and the traveling salesman problem. These three examples illustrate very well the mechanisms for incorporating problem-specific knowledge into specialized operators; in all examples operators transform feasible solutions into feasible offspring.

This is a very popular approach; most applications developed to date include some specialized operators which 'understand' the problem domain and preserve feasibility of solutions.

10.2 The transportation problem

We concentrate on operators which have been developed in connection with Genetic-2n, an evolutionary system for the transportation problem.

Three 'genetic' operators were defined: two mutations and one crossover. All these operators transform a matrix (representing a feasible transportation plan) into a new matrix (another feasible transportation plan). Note that a feasible matrix (i.e. a feasible transportation plan) should satisfy all marginal sums (i.e. totals for all rows and columns should be equal to given numbers, which represent supplies and demands at various sites). For example, let us assume that a transportation problem is defined with four sources and five destinations, where the supplies (vector sour) and demands (vector dest) are as follows:

$$\text{sour}[1] = 8.0 \quad \text{sour}[2] = 4.0 \quad \text{sour}[3] = 12.0 \quad \text{sour}[4] = 6.0$$

and

$$\text{dest}[1] = 3.0 \quad \text{dest}[2] = 5.0 \quad \text{dest}[3] = 10.0 \quad \text{dest}[4] = 7.0 \quad \text{dest}[5] = 5.0.$$

Then, the following matrix represents a feasible solution to the above transportation problem:

0.0	0.0	**5.0**	0.0	**3.0**
0.0	4.0	0.0	0.0	0.0
0.0	0.0	**5.0**	7.0	**0.0**
3.0	1.0	0.0	0.0	2.0

Note that the sum of all entries in the ith row is equal to sour[i] ($i = 1, 2, 3, 4$) and the total of all entries in the jth column is equal to dest[j] ($j = 1, 2, 3, 4, 5$).

The first mutation selects some (random) number of rows and columns from a parent matrix; assume that the first and the third rows were selected together with the first, third, and fifth columns. The entries which are placed on the intersection of selected rows and columns (typed in boldface in the original matrix) form the following submatrix:

0.0	5.0	3.0
0.0	5.0	0.0

In this submatrix all marginal sums are calculated, and all values are reinitialized. The initialization procedure introduces as many zero entries into the matrix as possible, thus searching the surface of the feasible convex search space. All marginal totals are left unchanged. For example, the following submatrix may result after the reinitialization process is completed:

0.0	8.0	0.0
0.0	2.0	3.0

Consequently, the offspring matrix (a new feasible transportation plan) is:

0.0	0.0	8.0	0.0	0.0
0.0	4.0	0.0	0.0	0.0
0.0	0.0	2.0	7.0	3.0
3.0	1.0	0.0	0.0	2.0

The only difference between the two mutation operators is that the second one avoids introducing zero while reinitializing submatrices, thus moving a feasible solution towards the center of the feasible search space.

The third operator, arithmetical crossover, for any two feasible parents (matrices U and V) produces two children X and Y, where $X = c_1 U + c_2 V$ and $Y = c_1 V + c_2 U$ (where $c_1, c_2 \geq 0$ and $c_1 + c_2 = 1$). As the constraint set is convex this operation ensures that both children are feasible if both parents are.

It is clear that all operators of Genetic-2n maintain feasibility of potential solutions: arithmetical crossover produces a point between two feasible points of the convex search space and both mutations were restricted to submatrices only to ensure no change in marginal sums.

10.3   Nonlinear optimization with linear constraints

Let us consider the following optimization problem: optimize a function $f(x_1, x_2, \ldots, x_n)$ subject to the following sets of linear constraints:

(i)   *Domain constraints.* $l_i \leq x_i \leq u_i$ for $i = 1, 2, \ldots, n$. We write $l \leq x \leq u$, where $l = (l_1, \ldots, l_n)$, $u = (u_1, \ldots, u_n)$, $x = (x_1, \ldots, x_n)$.

(ii) *Equalities.* $Ax = b$, where $x = (x_1, \ldots, x_n)$, $A = (a_{ij})$, $b = (b_1, \ldots, b_p)$, $1 \leq i \leq p$, and $1 \leq j \leq n$ (p is the number of equations).

(iii) *Inequalities.* $Cx \leq d$, where $x = (x_1, \ldots, x_n)$, $C = (c_{ij})$, $d = (d_1, \ldots, d_m)$, $1 \leq i \leq m$, and $1 \leq j \leq n$ (m is the number of inequalities).

Due to the linearity of the constraints, the solution space is always a convex space $\mathcal{D}$. Convexity of $\mathcal{D}$ implies that:

- for any two points s_1 and s_2 in the solution space $\mathcal{D}$, the linear combination $as_1 + (1 - a)s_2$, where $a \in [0, 1]$, is a point in $\mathcal{S}$.
- for every point $s_0 \in \mathcal{S}$ and any line p such that $s_0 \in p$, p intersects the boundaries of $\mathcal{S}$ at precisely two points, say $l_p^{s_0}$ and $u_p^{s_0}$.

Consequently, the value of the ith component of a feasible solution $x = (x_1, \ldots, x_n)$ is always in some (dynamic) range (left(i), right(i)); the bounds left(i) and right(i) depend on the other vector values $x_1, \ldots, x_{i-1}, x_{i+1}, \ldots, x_n$, and the set of constraints.

Several specialized operators were developed on the basis of the above properties; we discuss some of them in turn.

Uniform mutation. This operator requires a single parent x and produces a single offspring x'. The operator selects a random component $k \in \{1, \ldots, n\}$ of the vector $x = (x_1, \ldots, x_k, \ldots, x_n)$ and produces $x' = (x_1, \ldots, x'_k, \ldots, x_n)$, where x'_k is a random value (uniform probability distribution) from the range $(\text{left}(k), \text{right}(k))$.

Boundary mutation. This operator requires also a single parent x and produces a single offspring x'. The operator is a variation of the uniform mutation with x'_k being either $\text{left}(k)$ or $\text{right}(k)$, each with equal probability.

Nonuniform mutation. This is the (unary) operator responsible for the fine-tuning capabilities of the system. It is defined as follows. For a parent x, if the element x_k is selected for this mutation, the result is $x' = (x_1, \ldots, x'_k, \ldots, x_n)$, where

$$x'_k = \begin{cases} x_k + \triangle(t, \text{right}(k) - x_k) & \text{if a random binary digit is 0} \\ x_k - \triangle(t, x_k - \text{left}(k)) & \text{if a random binary digit is 1.} \end{cases}$$

The function $\triangle(t, y)$ returns a value in the range $[0, y]$ such that the probability of $\triangle(t, y)$ being close to zero increases as t increases (t is the generation number). This property causes this operator to search the space uniformly initially (when t is small), and very locally at later stages. We have used the following function:

$$\triangle(t, y) = yr(1 - t/T)^b$$

where r is a random number in the range $[0..1]$, T is the maximal generation number, and b is a system parameter determining the degree of nonuniformity.

Arithmetical crossover. This binary operator is defined as a linear combination of two vectors: if x_1 and x_2 are to be crossed, the resulting offspring are $x'_1 = ax_1 + (1 - a)x_2$ and $x'_2 = ax_2 + (1 - a)x_1$. This operator uses a random value $a \in [0..1]$, as it always guarantees closedness ($x'_1, x'_2 \in \mathcal{D}$).

All of the above operators preserve feasibility, transforming (a) feasible parent(s) into feasible offspring.

10.4   Traveling salesman problem

Whitley *et al* (1989) developed the *edge recombination* crossover (ER) for the traveling salesman problem. The ER operator explores the information on edges in a tour, for example for the tour

$$(3\ 1\ 2\ 8\ 7\ 4\ 6\ 9\ 5)$$

the edges are (3 1), (1 2), (2 8), (8 7), (7 4), (4 6), (6 9), (9 5), and (5 3). After all, edges—not cities—carry values (distances) in the TSP. The objective function to be minimized is the total of edges which constitute a legal tour. The position of a city in a tour is not important: tours are circular. Also, the direction of an edge is not important: both edges (3 1) and (1 3) signal only that cities 1 and 3 are directly connected.

The general idea behind the ER crossover is that an offspring should be built exclusively from the edges present in both parents. This is done with help of the edge list created from both parent tours. The edge list provides, for each city c, all other cities connected to city c in at least one of the parents. Obviously, for each city c there are at least two and at most four cities on the list. For example, for the two parents

$$p_1 = (1\ 2\ 3\ 4\ 5\ 6\ 7\ 8\ 9)$$

and

$$p_2 = (4\ 1\ 2\ 8\ 7\ 6\ 9\ 3\ 5)$$

the edge list is

city 1 : edges to other cities: 9 2 4
city 2 : edges to other cities: 1 3 8
city 3 : edges to other cities: 2 4 9 5
city 4 : edges to other cities: 3 5 1
city 5 : edges to other cities: 4 6 3
city 6 : edges to other cities: 5 7 9
city 7 : edges to other cities: 6 8
city 8 : edges to other cities: 7 9 2
city 9 : edges to other cities: 8 1 6 3.

The construction of the offspring starts with a selection of an initial city from one of the parents. Whitley *et al* (1989) selected one of the initial cities (e.g. 1 or 4 in the example above). The city with the smallest number of edges in the edge list is selected. If these numbers are equal, a random choice is made. Such selection increases the chance that we complete a tour with all edges selected from the parents. With a random selection, the chance of having edge failure, that is, being left with a city without a continuing edge, would be much higher. Assume we have selected city 1. This city is directly connected with three other cities: 9, 2, and 4. The next city is selected from these three. In our example, cities 4 and 2 have three edges, and city 9 has four. A random choice is made between cities 4 and 2; assume city 4 was selected. Again, the candidates for the next city in the constructed tour are 3 and 5, since they are directly connected to the last city, 4. Again, city 5 is selected, since it has only three edges as opposed to the four edges of city 3. So far, the offspring has the following shape:

$$(1\ 4\ 5\ x\ x\ x\ x\ x\ x).$$

Continuing this procedure we finish with the offspring

$$(1\ 4\ 5\ 6\ 7\ 8\ 2\ 3\ 9)$$

which is composed entirely of edges taken from the two parents.

The ER crossover was further enhanced (Starkweather *et al* 1991). The idea was that the 'common subsequences' were not preserved in the ER crossover. For example, if the edge list contains the row with three edges

city 4 : edges to other cities: 3 5 1

then one of these edges repeats itself. Referring to the previous example, it is the edge (4 5). This edge is present in both parents. However, it is listed as other edges, for example (4 3) and (4 1), which are present in one parent only. The proposed solution modifies the edge list by storing 'flagged' cities:

city 4 : edges to other cities: 3 -5 1.

The notation '-' means simply that the flagged city 5 should be listed twice. In the previous example of two parents

$$p_1 = (1\ 2\ 3\ 4\ 5\ 6\ 7\ 8\ 9)$$

and

$$p_2 = (4\ 1\ 2\ 8\ 7\ 6\ 9\ 3\ 5)$$

the (enhanced) edge list is:

city 4 : edges to other cities: 9 -2 4

city 4 : edges to other cities: -1 3 8

city 4 : edges to other cities: 2 4 9 5

city 4 : edges to other cities: 3 -5 1

city 4 : edges to other cities: -4 6 3

city 4 : edges to other cities: 5 -7 9

city 4 : edges to other cities: -6 -8

city 4 : edges to other cities: -7 9 2

city 4 : edges to other cities: 8 1 6 3.

The algorithm for constructing a new offspring gives priority to flagged entries: this is important only in the cases where three edges are listed—in the two other cases either there are no flagged cities, or both cities are flagged. This enhancement (plus a modification for making better choices when random edge selection is necessary) further improved the performance of the system.

We illustrate this enhancement using the previous example. Assume we have selected city 1 as an initial city for an offspring. As before, this city is directly connected with three other cities: 9, 2, and 4. In this case, however, city 2 is flagged, so it is selected as the next city of the tour. Again, the candidates for the next city in the constructed tour are 3 and 8, since they are directly connected to the last city, 2; the flagged city 1 is already present in the partial tour and is not considered. Again, city 8 is selected, since it has only three edges as opposed to the four edges of city 3. So far, the offspring has the following shape:

$$(1\ 2\ 8\ x\ x\ x\ x\ x\ x).$$

Continuing this procedure we finish with the offspring

$$(1\ 2\ 8\ 7\ 6\ 5\ 4\ 3\ 9)$$

which is composed entirely of edges taken from the two parents.

References

Davis L (ed) 1987 *Genetic Algorithms and Simulated Annealing* (Los Altos, CA: Morgan Kaufmann)
——1991 *Handbook of Genetic Algorithms* (New York: Van Nostrand Reinhold)
Radcliffe N J 1991 Forma analysis and random respectful recombination *Proc. 4th Int. Conf. on Genetic Algorithms (San Diego, CA, July 1991)* ed R Belew and L Booker (San Mateo, CA: Morgan Kaufmann) pp 222–9
Radcliffe N J 1994 The algebra of genetic algorithms *Ann. Maths Artificial Intell.* **10** 339–84
Starkweather T, McDaniel S, Mathias K, Whitley C and Whitley D 1991 A comparison of genetic sequencing operators *Proc. 4th Int. Conf. on Genetic Algorithms (San Diego, CA, July 1991)* ed R Belew and L Booker (San Mateo, CA: Morgan Kaufmann) pp 69–76
Surry P D, Radcliffe N J and Boyd I D 1995 A multi-objective approach to constrained optimization of gas supply networks *AISB-95 Workshop on Evolutionary Computing (Sheffield, 1995)* ed T Fogarty (Berlin: Springer) pp 166-80
Whitley D, Starkweather T and Fuquay D'A 1989 Scheduling problems and traveling salesman: the genetic edge recombination operator *Proc. 3rd Int. Conf. on Genetic Algorithms (Fairfax, VA, June 1989)* ed J D Schaffer (San Mateo, CA: Morgan Kaufmann) pp 133–40

11

Other constraint-handling methods

Zbigniew Michalewicz

11.1 Introduction

Several additional constraint-handling heuristics have emerged during the last few years. Often these methods are difficult to classify: either they are based on some new ideas or they combine a few elements present in other methods. In this section several such techniques are discussed.

11.2 Multiobjective optimization methods

One of the techniques includes utilization of multiobjective optimization methods, where the objective function f and, for m constraints

$$g_j(x) \leq 0 \qquad \text{for } j = 1, \ldots, q$$

and

$$h_j(x) = 0 \qquad \text{for } j = q + 1, \ldots, m$$

their constraint violation measures f_j

$$f_j(x) = \begin{cases} \max\{0, g_j(x)\} & \text{if } 1 \leq j \leq q \\ |h_j(x)| & \text{if } q + 1 \leq j \leq m \end{cases}$$

constitute an $(m + 1)$-dimensional vector:

$$\text{eval}(x) = (f, f_1, \ldots, f_m).$$

Using some multiobjective optimization method, we can attempt to minimize its components: an ideal solution x would have $f_j(x) = 0$ for $1 \leq i \leq m$ and $f(x) \leq f(y)$ for all feasible y (minimization problems).

A successful implementation of a similar approach was presented recently by Surry *et al* (1995). All individuals in the population are measured with respect to constraint satisfaction—each individual x is assigned a rank $r(x)$ according to its Pareto ranking; the rank can be assigned either by peeling off successive nondominating layers or by calculating the number of solutions which dominate it (see Chapter 5 for more details on multiobjective optimization). Then, the evaluation measure of each individual is given as a two-dimensional vector:

$$\text{eval}(x) = \langle f(x), r(x) \rangle.$$

At this stage, a modified Schaffer (1984) VEGA system (for vector evaluated genetic algorithm) was used. The main idea behind the VEGA system was a division of the population into (equal-sized) subpopulations; each subpopulation was 'responsible' for a single objective. The selection procedure was performed independently for each objective, but crossover was performed across subpopulation boundaries. Additional heuristics were developed (e.g. wealth redistribution scheme, crossbreeding plan) and studied to decrease a tendency of the system to converge towards individuals which were not the best with respect to any objective. However, instead of proportional selection, a binary tournament selection was used, where the tournament criterion cost value f is selected with probability p and constraint ranking r with probability $1 - p$. The value of the parameter p is adapted during the run of the algorithm; it is increased or decreased on the basis of the ratio of feasible individuals in the recent generations (i.e. if the ratio is too low, the parameter p is decreased. Clearly, if p approaches zero, the system favors constraint rank). The outline of the system implemented for a particular problem of the optimization of gas supply networks (Surry *et al* 1995) is as follows:

- calculate constraint violation for all solutions
- rank individuals based on constraint violation (Pareto ranking)
- evaluate the cost of solutions (in terms of function f)
- select proportion p of parents based on cost, and the other based on ranking
- perform genetic operators
- adjust p on the basis of the ratio of feasible individuals in the recent generations.

11.3   Coevolutionary model approach

Another approach was reported by Paredis (1994). The method (described in the context of constraint satisfaction problems) is based on a coevolutionary model, where a population of potential solutions coevolves with a population of constraints: fitter solutions satisfy more constraints, whereas fitter constraints are violated by more solutions (i.e. the harder it is to satisfy a constraint, the fitter the constraint is, thus participating more actively in evaluating individuals from the solution space). This means that individuals from the population of solutions are considered from the whole search space S, and that there is no distinction between feasible and infeasible individuals (i.e. there is only one evaluation function eval without any split into eval_f for feasible or eval_u for infeasible individuals). The value of eval is determined on the basis of constraint violation measures f_j; however, fitter constraints (e.g. active constraints) would contribute more frequently to the value of eval.

Yet another heuristic is based on the idea of handling constraints in a particular order; Schoenauer and Xanthakis (1993) called this method a 'behavioral memory' approach. The initial steps of the method are devoted to sampling the feasible region; only in the final step is the objective function f optimized.

- Start with a random population of individuals (i.e. these individuals are feasible or infeasible).
- Set $j = 1$ (j is the constraint counter).
- Evolve this population to minimize the violation of the jth constraint, until a given percentage of the population (the so-called flip threshold ϕ) is feasible for this constraint. In this case

$$\text{eval}(x) = g_1(x).$$

- Set $j = j + 1$.
- The current population is the starting point for the next phase of the evolution, minimizing the violation of the jth constraint:

$$\text{eval}(x) = g_j(x).$$

(To simplify notation, we do not distinguish between inequality constraints g_j and equations h_j; all m constraints are denoted by g_j.) During this phase, points that do not satisfy at least one of the first, second, ..., $(j-1)$th constraints are eliminated from the population. The halting criterion is again the satisfaction of the jth constraint by the flip threshold percentage ϕ of the population.

- If $j < m$, repeat the last two steps, otherwise ($j = m$) optimize the objective function f rejecting infeasible individuals.

The method has a few merits. One of them is that in the final step of the algorithm the objective function f is optimized (as opposed to its modified form). However, for larger feasible spaces the method just provides additional computational overhead, and for very small feasible search spaces it is essential to maintain diversity in the population.

11.4 Cultural algorithms

It is also possible to incorporate the knowledge of the constraints of the problem into the belief space of cultural algorithms (Reynolds 1994); such algorithms provide a possibility of conducting an efficient search of the feasible search space (Reynolds *et al* 1995). The research on cultural algorithms (Reynolds 1994) was triggered by observations that culture might be another kind of inheritance system. However it is not clear what the appropriate structures and units to represent the adaptation and transmission of cultural information are. Neither is it clear how to describe the interaction between natural evolution and culture. Reynolds developed a few models to investigate the properties of cultural algorithms; in these models, the belief space is used to constrain the combination of traits that individuals can assume. Changes in the belief space represent macroevolutionary change and changes in the population of individuals represent microevolutionary change. Both changes are moderated by the communication link.

The general intuition behind belief spaces is to preserve those beliefs associated with 'acceptable' behavior at the trait level (and, consequently, to prune away unacceptable beliefs). The acceptable beliefs serve as constraints that direct the population of traits. It seems that the cultural algorithms may serve as a very interesting tool for numerical optimization problems, where constraints influence the search in a direct way (consequently, the search may be more efficient in constrained spaces than in unconstrained ones!).

11.5 Segregated genetic algorithm

Le Riche *et al* (1995) proposed a method which combines the ideas of penalizing infeasible solutions with coevolutionary concepts. The 'classical' methods based on penalty functions either (i) maintain static penalty coefficients (see e.g. Homaifar *et al* 1994), (ii) use dynamic penalties (Smith and Tate 1993), (iii) use penalties which are functions of the evolution time (Michalewicz and Attia 1994, Joines and Houck 1994), or (iv) adapt penalty coefficients on the basis of the number of feasible and infeasible individuals in recent generations (Bean and Hadj-Alouane 1992). The method of Le Riche proposes a so-called segregated genetic algorithm which uses a double-penalty strategy. The population is split into two coevolving subpopulations, where the fitness of each subpopulation is evaluated using either one of the two penalty parameters. The two subpopulations converge along two complementary trajectories, which may help to locate the optimal region faster. Also, such a system may make the algorithm less sensitive to the choice of penalty parameters.

The outline of the method is as follows:

- create two sets of penalty coefficients, p_j and r_j ($j = 1, \ldots, m$), where $p_i \ll r_i$
- start with two random populations of individuals (i.e. these individuals are feasible or infeasible); the size of each population is the same pop_size
- evaluate these populations; the first population is evaluated as

$$\text{eval}(\boldsymbol{x}) = f(\boldsymbol{x}) + \sum_{j=1}^{m} p_j f_j^2(\boldsymbol{x})$$

and the other population as

$$\text{eval}(\boldsymbol{x}) = f(\boldsymbol{x}) + \sum_{j=1}^{m} r_j f_j^2(\boldsymbol{x})$$

- create two separate ranked lists
- merge the two lists into one ranked population of the size pop_size
- apply selection and operators to the new population; create the new population of pop_size offspring
- evaluate the new population twice (with respect to p_j and r_j values, respectively)
- from the old and new populations create two populations of the size pop_size each; each population is ranked accordingly to its evaluation with respect to p_j and r_j values, respectively
- repeat the last four steps.

The major merit of this method is that it permits the balancing of the influence of two sets of penalty parameters. Note that if $p_j = r_j$ (for $1 \le j \le m$), the algorithm is similar to the (pop_size, pop_size) evolution strategy. The reported results of applying the above segregated genetic algorithm to the laminate design problem were very good (Le Riche *et al* 1995).

11.6 Genocop III

Michalewicz and Nazhiyath (1995) mixed the idea of repair algorithms with concepts of coevolution. The Genocop III system maintains two separate populations: the first population consists of (not necessarily feasible) search points and the second population consists of fully feasible reference points. Reference points, being feasible, are evaluated directly by the objective function. Search points are 'repaired' for evaluation. The repair process samples the segment between the search and reference points; the first feasible point is accepted as a repaired version of the search point. Genocop III avoids some disadvantages of other systems. It uses the objective function for evaluation of fully feasible individuals only, so the evaluation function is not distorted as in methods based on penalty functions. It introduces only few additional parameters and it always returns a feasible solution. However, it requires an efficient repair process, which might be too costly for many engineering problems.

A comparison of some of the above methods is presented by Michalewicz (1995).

References

Bean J C and Hadj-Alouane A B 1992 *A Dual Genetic Algorithm for Bounded Integer Programs* Department of Industrial and Operations Engineering, University of Michigan, TR 92-53

Hadj-Alouane A B and Bean J C 1992 *A Genetic Algorithm for the Multiple-Choice Integer Program* Department of Industrial and Operations Engineering, University of Michigan, TR 92-50

Homaifar A, Lai S H-Y and Qi X 1994 Constrained optimization via genetic algorithms *Simulation* **62** 242–54

Juliff K 1993 A multi-chromosome genetic algorithm for pallet loading *Proc. 5th Int. Conf. on Genetic Algorithms* ed S Forrest (San Mateo, CA: Morgan Kaufmann) pp 467–73

Le Riche R G, Knopf-Lenoir C and Haftka R T 1995 A segregated genetic algorithm for constrained structural optimization *Proc. 6th Int. Conf. on Genetic Algorithms* ed L Eshelman (San Mateo, CA: Morgan Kaufmann) pp 558–65

Michalewicz Z 1995 Genetic algorithms, numerical optimization and constraints *Proc. 6th Int. Conf. on Genetic Algorithms* ed L Eshelman (San Mateo, CA: Morgan Kaufmann) pp 151–8

Michalewicz Z and Attia N 1994 Evolutionary optimization of constrained problems *Proc. 3rd Ann. Conf. on Evolutionary Programming* ed A V Sebald and L J Fogel (Singapore: World Scientific) pp 98–108

Michalewicz Z and Nazhiyath G 1995 Genocop III: a co-evolutionary algorithm for numerical optimization problems with nonlinear constraints *Proc. 2nd IEEE Int. Conf. on Evolutionary Computation (Perth, 1995)* pp 647–51

Paredis J 1994 Co-evolutionary constraint satisfaction *Proc. 3rd Conf. on Parallel Problem Solving from Nature* (New York: Springer) pp 46–55

Reynolds R G 1994 An introduction to cultural algorithms *Proc. 3rd Ann. Conf. on Evolutionary Programming* ed A V Sebald and L J Fogel (Singapore: World Scientific) pp 131–9

Reynolds R G, Michalewicz Z and Cavaretta M 1995 Using cultural algorithms for constraint handling in Genocop *Proc. 4th Ann. Conf. on Evolutionary Programming (San Diego, CA, 1995)* ed J R McDonnell, R G Reynolds and D B Fogel pp 289–305

Schaffer J D 1984 *Some Experiments in Machine Learning Using Vector Evaluated Genetic Algorithms* PhD Dissertation, Vanderbilt University

Schoenauer M and Xanthakis S 1993 Constrained GA optimization *Proc. 5th Int. Conf. on Genetic Algorithms* ed S Forrest (Los Altos, CA: Morgan Kaufmann) pp 573–80

Smith A E and Tate D M 1993 Genetic optimization using a penalty function *Proc. 5th Int. Conf. on Genetic Algorithms* ed S Forrest (Los Altos, CA: Morgan Kaufmann) pp 499–503

Surry P D, Radcliffe N J and Boyd I D 1995 A multi-objective approach to constrained optimization of gas supply networks *AISB-95 Workshop on Evolutionary Computing (Sheffield, 1995)*

12

Constraint-satisfaction problems

A E Eiben and Zs Ruttkay

12.1 Introduction

Applying evolutionary algorithms (EAs) for solving constraint-satisfaction problems (CSP) is interesting from two points of view. On the one hand, since a general CSP is known to be NP complete (Mackworth 1977), one cannot expect that an effective classical deterministic search algorithm can be forged to solve CSPs. There has been a continuous effort to construct effective algorithms for specific CSPs, and to characterize the difficulty of problem classes. The proposed algorithms apply different search strategies, often guided by heuristics based on some evaluation of the uninstantiated variables and of the possible values. The search can be preceded by preprocessing of the domains or the constraints. For an overview of specific search strategies and heuristics see the work of Meseguer (1989), Nudel (1983), and Tsang (1993). What makes deterministic heuristic search methods strong in certain cases is just what makes them weak in others: they restrict the scope of the search, based on (explicit or implicit) heuristics. If the heuristics turn out to be misleading, it is often very tiresome to enlarge or shift the scope of the search using a series of backtrackings. This problem could be treated by diversifying the search by maintaining several different candidate solutions in parallel and counterbalancing the greediness of the heuristics by incorporating random elements into the construction mechanism of new candidates. These two principles are essential for EAs. Hence, the idea of applying EAs to solve CSPs is a natural response for the limitations of the classical CSP solving methods.

On the other hand, traditional EAs are mainly used for unconstrained optimization. Problems where constraints play an essential role have the common reputation of being EA hard. This is due to the fact that standard genetic operators (mutation and crossover) are 'blind' to constraints. In other words, there is no guarantee that children of feasible parents are also feasible, nor that children of infeasible parents are 'less infeasible' than the parents. Thus, handling constrained problems with EAs is a big challenge for the field (cf Michalewicz and Michalewicz 1995).

The main goals of this section are to present definitions that yield a clear conceptual framework and terminology for constrained problems and to discuss different ways to apply EAs for solving CSPs within the above framework.

12.2 Free optimization, constrained optimization, and constraint satisfaction

Let us first set up a general conceptual framework for constrained problems. Without such a framework terminology and views can be ambiguous. For instance, the traveling salesperson problem (TSP) is a constrained problem if we consider that each variable can have each city label as a value but a solution can contain each label only once. This latter restriction is a constraint on the search space $S = D_1 \times \ldots \times D_n$, where each D_i ($i \in \{1, \ldots, n\}$) is the set of all city labels. Nevertheless, the TSP is an unconstrained problem if we define the search space as the set of all permutations of the city labels.

Definition 12.2.1. We will call a Cartesian product of sets $S = D_1 \times \ldots \times D_n$ a *free search space*.

Note that this definition puts no requirements on the domains; they can be discrete or continuous, connected or not. The rationale behind this definition is that testing the membership relation of a free search space can be performed independently on each coordinate and taking the conjunction of the results. This implies an interesting property from an EA point of view: if two chromosomes from a free search space are crossed over, their offspring will be in the same space as well. Thus, (genetic) search in a space of the form $S = D_1 \times \ldots \times D_n$ is free in this sense; this motivates the name.

Definition 12.2.2. A *free optimization problem* (FOP) is a pair $\langle S, f \rangle$, where S is a free search space and f is a (real-valued) objective function on S, which has to be optimized (minimized or maximized). A *solution of a free optimization problem* is an $s \in S$ with an optimal f-value.

Definition 12.2.3. A *constraint-satisfaction problem* (CSP) is a pair $\langle S, \phi \rangle$, where S is a free search space and ϕ is a formula (a Boolean function on S). A *solution of a constraint-satisfaction problem* is an $s \in S$ with $\phi(s) = true$.

Usually a CSP is stated as a problem of finding an instantiation of variables $v_1, \ldots, v_n$ within the finite domains $D_1, \ldots, D_n$ such that constraints (relations) $c_1, \ldots, c_m$ prescribed for (some of) the variables hold. The formula ϕ is then the conjunction of the given constraints. One may be interested in one, some, or all solutions, or only in the existence of a solution. We restrict our discussion to finding one solution. It is also worth noting that our definitions allow CSPs with continuous domains. Such a case is almost never considered in the CSP literature: by the finiteness assumption on the domains $D_1, \ldots, D_n$ the usual CSPs are discrete.

Definition 12.2.4. A *constrained optimization problem* (COP) is a triple $\langle S, f, \phi \rangle$, where S is a free search space, f is a (real-valued) objective function on S and ϕ is a formula (a Boolean function on S). A *solution of a constrained optimization problem* is an $s \in S$ with $\phi(s) = true$ and an optimal f-value.

The above three problem types can be represented in the same scheme as $\langle S, f, \bullet \rangle$, $\langle S, \bullet, \phi \rangle$, and $\langle S, f, \phi \rangle$ respectively, where $\bullet$ means the absence of the given component.

Definition 12.2.5. For CSPs, as well as for COPs, we call the ϕ the *feasibility condition*, and the set $\{s \in S | \phi(s) = true\}$ will be called the *feasible search space*.

With this terminology, solving a CSP means finding one single feasible element; solving a COP means finding a feasible and optimal element. These definitions eliminate the arbitrariness of viewing a problem as a constrained problem. For example, in the most natural formalization of the TSP candidate solutions are permutations of the cities $\{x_1, \ldots, x_n\}$. The TSP is then a COP $\langle S, f, \phi \rangle$, where $S = \{x_1, \ldots, x_n\}^n$, $\phi(s) = true \Leftrightarrow \forall i, j \in \{1, \ldots, n\}\ s_i \neq s_j$ and $f(s) = \sum_{i=1}^{n} \text{dist}(s_i, s_{i+1})$, where $s_{n+1} := s_1$.

12.3 Transforming constraint-satisfaction problems to evolutionary-algorithm-suited problems

Note that the presence of an objective function (fitness function) to be optimized is essential for EAs. A CSP $\langle S, \bullet, \phi \rangle$ is lacking this component; in this sense it is not EA suited. Therefore it needs to be transformed to an EA-suited problem, an FOP $\langle S, f, \bullet \rangle$ or a COP $\langle S, f, \psi \rangle$, before an EA can be applied to it.

Definition 12.3.1. Let problems A and B be either of $\langle S, f, \bullet \rangle$, $\langle S, \bullet, \phi \rangle$, $\langle S, f, \phi \rangle$. A and B are *equivalent* if

$$\forall s \in S : s \text{ is a solution of } A \Longleftrightarrow s \text{ is a solution of } B.$$

We say that A *subsumes* B if

$$\forall s \in S : s \text{ is a solution of } A \Longrightarrow s \text{ is a solution of } B.$$

Thus, solving a CSP by an EA means that we transform it to an FOP/COP that subsumes it and solve this FOP/COP. In the following we discuss how to transform CSPs to FOPs and COPs. Solving the resulting problems is another issue. FOPs allow free search; in this sense they are simple for an EA; COPs, however, are difficult to solve by EAs. Notice that the term 'constraint handling' has two meanings in this context. Its first meaning is 'how to transform the constraints of a given CSP': whether, and how, to incorporate them in an FOP or a COP. The second meaning emerges when a CSP $\longrightarrow$ COP transformation is chosen: 'how to maintain the constraints when solving a COP by an EA'. It is this second meaning that is mostly intended in the EA literature.

12.3.1 Transforming a constraint-satisfaction problem to a free optimization problem

Let $\langle S, \bullet, \phi \rangle$ be a CSP and let us assume that ϕ is given by a conjunction of some constraints (relations) $c_1, \ldots, c_m$ that have to hold for the variables. An equivalent FOP can be created by defining an objective function f which has an optimal value if and only if all constraints $c_1, \ldots c_m$ are satisfied. Applying an EA for this FOP means that all constraints are handled *indirectly*, that is the EA operates freely on S (see the remark after definition 12.1) and reaching the optimum means satisfying all constraints. Using such a CSP $\longrightarrow$ FOP transformation implies that 'constraint handling' is restricted to its first meaning.

Incorporating all constraints in f can be done by applying penalties on constraint violation. The most straightforward possibility for a penalty function f based on the constraints is to consider the number of violated constraints. This measure, however, does not distinguish between difficult and easy constraints. These aspects can be reflected by assigning weights to the constraints and defining f as

$$f(s) = \sum_{i=1}^{m} w_i \times \chi(s, c_i) \tag{12.1}$$

where w_i is the penalty (or weight) assigned to constraint c_i and

$$\chi(s, c_i) = \begin{cases} 1 & \text{if } s \text{ violates } c_i \\ 0 & \text{otherwise.} \end{cases}$$

Satisfying a constraint with a high penalty gives a relatively high reward to the EA, hence it will be 'more interested' in satisfying such constraints. Thus, the definition of an appropriate penalty function (Chapter 7) is of crucial importance. For determining the constraint weights one can use the measures common in classical CSP solving methods (e.g. constraint tightness) to evaluate the difficulty of constraints. A more sophisticated notion of penalties can be based on the degree of violation for each constraint. In this approach $\chi(s, c_i)$ is not simply 1 or 0, but a measure of how severe the violation of c_i is.

Another type of penalty function is obtained if we concentrate on the variables, instead of the constraints. The function f is then based on the evaluation of the incorrect values, that is variables where the value violates at least one constraint. Let C^i ($i \in \{1, \ldots, n\}$) be the set of constraints that involves variable i. Then

$$f(s) = \sum_{i=1}^{n} w_i \times \chi(s, C^i) \tag{12.2}$$

where w_i is the penalty (or weight) assigned to variable i and

$$\chi(s, C^i) = \begin{cases} 1 & \text{if } s \text{ violates at least one } c \in C^i \\ 0 & \text{otherwise.} \end{cases}$$

Obviously, this approach can also be refined by measuring how serious the violation of constraints is. A particular implementation of this idea is to define $\chi(s, C^i)$ as $\sum_{c_j \in C^i} \chi(s, c_j)$.

Example 12.3.1. Consider the graph three-coloring problem, where the nodes of a given graph $G = (N, E)$, $E \subseteq N \times N$, have to be colored by three colors in such a way that no neighboring nodes, i.e. nodes connected by an edge, have the same color. Formally we can represent this problem as a CSP with $n = |N|$ variables, each with the same domain $D = \{1, 2, 3\}$. Furthermore, we need $m = |E|$ constraints, each belonging to one edge, with $c_e(s) = true$ iff $e = (k, l)$ and $s_k \neq s_l$, i.e. the two nodes on the corresponding edge have a different color. Then the corresponding CSP is $\langle S, \phi \rangle$, where $S = D^n$ and $\phi(s) = \bigwedge_{e \in E} c_e$. Using a constraint-oriented penalty function (equation (12.1)) with $w_e \equiv 1$ we would count the incorrect edges that connect two nodes with the same color. The variable-oriented penalty function (equation (12.2)) with $w_i \equiv 1$ amounts to counting the incorrect nodes that have a neighbor with the same color.

A great many penalty functions used in practice are (a variant of) one of the above two options. There are, however, other possibilities. For instance, instead of viewing the objective function f as a penalty for constraint violation, it can be perceived as the distance to a solution, or more generally, the cost of reaching a solution. In order to define such an f a distance measure d on the search space has to be given. Since the solutions are not known in advance, the real distance from $s \in S$ to the set of solutions can only be estimated. One such an estimation is based on the *projection of a constraint* c. If c operates on the variables $v_{i_1}, \ldots, v_{i_k}$, then this projection is defined as the set $S_c = \{\langle s_{i_1}, \ldots, s_{i_k} \rangle \in D_{i_1} \times \ldots \times D_{i_k} \mid c(s_{i_1}, \ldots, s_{i_k}) = true\}$, and the distance of $s \in S$ from S_c is $d(s, S_c) := \min\{d(\langle s_{i_1}, \ldots, s_{i_k} \rangle, z) \mid z \in S_c\}$. (We assume that d is also defined on hyperplanes of S.) Now it is a natural idea to estimate the distance of an $s \in S$ from a solution as

$$f(s) = \sum_{i=1}^{m} w_i \times d(s, S_{c_i}). \tag{12.3}$$

It is clear that $s \in S$ is a solution of the CSP iff $f(s) = 0$. Function (12.3) is just an example satisfying this equivalence property. In general an objective function of the third kind does not necessarily have to be based on a distance measure d. It can be any cost measure as long as $f(s) = 0$ implies membership of the set of solutions.

Example 12.3.2. Consider the graph three-coloring problem again. The projection of a constraint $c_{(k,l)}$ belonging to an edge (k, l) is $S_{c_{(k,l)}} = \{\langle 1, 2 \rangle, \langle 1, 3 \rangle, \langle 2, 1 \rangle, \langle 2, 3 \rangle, \langle 3, 1 \rangle, \langle 3, 2 \rangle\}$ and we can define the cost of reaching $S_{c_{(k,l)}}$ from an $s \in S$ as the number of value modifications in s needed to have $\langle s'_k, s'_l \rangle \in S_{c_{(k,l)}}$. In this simple example this will be one for every $s \in S$ and $c_{(k,l)}$, thus the formulas (12.1) and (12.3) will coincide, both counting the incorrect edges.

As we have already mentioned, solving an FOP is 'easy' for an EA; at least it is 'only' a matter of optimization. Whether or not the FOP approach is successful depends on the EA's ability to find an optimum. Penalizing infeasible individuals is studied extensively by Michalewicz (1995), primarily in the context of treating COPs with continuous domains. Experiments reported for example by Richardson *et al* (1989) and Michalewicz (1996, p 86–7) indicate that GAs with penalty functions are likely to fail on sparse problems, i.e. on problems where the ratio between the size of the feasible and the whole search space is small.

12.3.2 *Transforming a constraint-satisfaction problem to a constrained optimization problem*

The limitations of using penalty functions as the only means of handling constraints force one to look for other options. The basic idea is to incorporate only some of the constraints in f (these are handled indirectly) and maintain the other ones *directly*. This means that the CSP $\langle S, \bullet, \phi \rangle$ is transformed to a COP $\langle S, f, \psi \rangle$, where the constraints not in f form ψ. In this case we presume that the EA works with individuals satisfying ψ, that is, it will operate on the space $\{x \in S | \psi(x) = true\}$ and finding an $s \in \{x \in S | \phi(x) = true\}$ (a solution for the CSP) means finding an $s \in \{x \in S | \psi(x) = true\}$ with an optimal f-value.

Definition 12.3.2. If the context requires a clear distinction between ϕ (expressing the constraints given in the original CSP) and ψ (expressing an appropriate subset of the constraints in the COP) we will call ψ the *allowability condition* and $\{s \in S | \psi(s) = true\}$ the *allowable search space*.

For a given CSP several equivalent COPs can be defined by choosing the subset of the constraints incorporated in the allowability condition and/or defining the objective function measuring the satisfaction of the remaining constraints differently. Such decisions can be based on distiguishing constraints for which finding and maintaining solutions is easy (in ψ) or difficult (in f) (cf Smith and Tate 1993). For the constraints in the allowability condition it is essential that they can be satisfied by the initialization procedure and can be maintained by the EA. The latter requires that the EA guarantees that the new candidate solutions are always allowable. This implies that the COP approach is more complex than the FOP approach. When deciding which constraints to incorporate in the allowability condition, one should already consider how it can be maintained by the EA in mind.

12.3.3 Changing the search space

Up to now we have assumed that the candidate solutions of the original CSP and the individuals of the EA are of the same type, namely members of S. This is not necessary: the EA may operate on a different search space, S', assuming that a decoder (Chapter 8) is given which transforms a given genotype $s' \in S'$ into a phenotype $s \in S$. Usually, this technique is used in such a way that the elements of S generated by decoding elements of S' automatically fulfill some of the original constraints. However, it is not guaranteed automatically that the decoder can create the whole S from S', hence, it can occur that not all solutions of the original CSP can be produced.

Here again, choosing an appropriate representation, i.e. S', for the EA and making the decoder are highly correlated. Designing a good decoder is clearly a problem specific task. However, order-based representation, where S' consists of permutations, is a generally advisable option. Many decoders create a search object (an $s \in S$) by a sequential procedure, following a certain order of processing. In these terms the goal of the search is to find a sequence that encodes a solution; that is, we have a sequencing problem. In the meanwhile, sequencing problems can be naturally represented by permutations as chromosomes and there are many off-the-shelf order-based operators at our disposal (Olivier *et al* 1987, Fox and McMahon 1991, Starkweather *et al* 1991). This makes order-based representation a promising option.

Example 12.3.3. For the graph three-coloring problem each permutation of nodes can be assigned a coloring by a simple greedy decoding algorithm. The decoder processes the nodes in the order they appear in the permutation π and colors node π_i with the smallest color from $\{1, 2, 3\}$ that does not violate the constraints. If none of the colors in $\{1, 2, 3\}$ is suitable a random assignment is made. Formally, we change to a COP $\langle S', f, \psi \rangle$, where $S' = \{s_1, \ldots, s_n\}^n$, $\psi(s') = true \Leftrightarrow \forall i, j \in \{1, \ldots, n\}$ $s_i' \neq s_j'$ and f is an objective function taking its optimum on permutations that encode feasible colorings. Note that when coloring a node π_i some neighbors of i might not have a color yet, thus not all constraints can be evaluated. This means that the decoder considers a color suitable if it does not violate that subset of the constraints that can be evaluated at the given moment.

An interesting variation of this permutations + decoder approach is decoding permutations to *partial* solutions. In the work of Eiben *et al* (1994) and Eiben and van der Hauw (1996) the decoder leaves nodes uncolored in the case of violations and f is the number of uncolored nodes. It might seem that this objective function supplies too little information, but the experiments showed that this EA consistently outperforms the one with integer representation (see example 12.1). This is even more interesting if we take into account that the permutation space has $n!$ elements, while the size of $S = \{1, 2, 3\}^n$ is only 3^n.

12.4 Solving the transformed problem

If we have transformed a given CSP to an FOP we can simply apply the usual
EA machinery to find a solution of this FOP. However, there can be many ways
to enhance a simple function optimizing EA by incorporating constraint-specific
features into it. Next we discuss two domain independent options, concerning
the search operators, respectively the fitness function. One option is thus to
use specific search operators based on heuristics that try to create children that
are 'less infeasible' than the parents. The heuristic operators used by Eiben
et al (1994, 1995) work by (partially) replacing pure random mutation and
recombination mechanisms by heuristic ones. Domain independent variable- and
value-ordering heuristics from the classical constructive CSP solving methods
are adopted. There are two kinds of heuristic, one for selecting the position
to change in a chromosome and one for choosing a new value for a selected
variable. The heuristic for selecting the position to change chooses that gene
i which causes the most severe violation in terms of $\sum_{c_j \in C^i} \chi(s, c_j)$. The
heuristic for value selection choose a value that leads to the highest decrease
in penalty. Using these problem independent techniques the performance of
genetic algorithms (GAs) on CSPs can be highly increased.

Another option is to dynamically refocus the search by changing the
objective function f. The basic idea is that the weights are periodically modified
by an adaptive mechanism depending on the progress of the search. If in the
best individual a constraint is not satisfied, or a variable is instantiated to a value
that violates constraints, then the corresponding weight is increased. We have
applied this approach successfully for solving CSPs. We observed (Eiben and
Ruttkay 1996) that the GA was able to learn constraint weights that were to a
large extent independent from the applied genetic operators and initial constraint
weights. We showed (Eiben and van der Hauw 1996) that this technique is very
powerful and it resulted in a superior graph-coloring algorithm. A big advantage
of this technique is that it is problem independent. Coevolutionary constraint
satisfaction (Paredis 1994) can also be seen as a particular implementation
of adapting the penalty function. Dynamically changing penalties were also
applied by Michalewicz and Attia (1994) and Smith and Tate (1993), for solving
(continuous) COPs.

If we have chosen to transform the original CSP to a COP (whether or not
through a decoder), then we have to take care to enforce the constraints in the
allowability condition. This is typically done by either:

(i) *eliminating* newborn individuals if they are not allowable,
(ii) *preserving* allowability, i.e. using special operators that guarantee that
    allowable parents have allowable children, or
(iii) *repairing* newborn individuals if they are not allowable.

The eliminating approach is generally very inefficient, therefore hardy
practicable. Repair algorithms are treated in Section C5.4(Chapter 9), while
Chapter 10 handles constraint-preserving operators; therefore we omit a detailed

discussion here. Let us, however, make a remark on the preserving approach. When choosing the subset of the constraints incorporated in ψ or in f one should keep in mind that the more constraints are represented by f, the less informative the evaluation of the individuals is. For instance, we know that two constraints are violated, but we do not know which ones. Hence, the performance of an EA can be expected to be raised by putting many constraints in ψ and few constraints in f. Nevertheless, this requires genetic operators that maintain many constraints. Thus, the representation issue, i.e. what constraints to keep in the allowability criterion, cannot be treated independently from the operator issue (see also De Jong 1987).

12.5 Conclusions

In this section we defined three problem classes, FOPs, CSPs, and COPs. In this framework we gave a systematic overview of possibilities to apply EAs for CSPs. Since a CSP has no optimization component it has to be transformed to an FOP or a COP that subsumes it and an EA should be applied to the transformed problem. The FOP option means unconstrained search, thus success of this approach depends on the ability of minimizing penalties. To this end we have sketched two problem independent extensions to a general evolutionary optimizer:

- applying heuristic operators based on classical CSP solving techniques that presumably reduce the level of constraint violation, and
- using adaptive penalty functions based on an updating mechanism of weights, thus allowing the EA to redefine the relative importance of constraints or variables to focus on.

Once we have a COP to be solved by an EA the constraints in the feasibility (allowability) conditions have to be taken care of, while still having to minimize penalties. Solving COPs by EAs has already received a lot of attention: for detailed discussions we refer to other sections of this handbook. Let us note, however, that the EA extensions mentioned for the FOP-based approach are also applicable for improving the performance of an EA working on a COP.

The quoted examples and other successful case studies (e.g. Dozier *et al* 1994, Hao and Dorne 1994) suggest that for many CSPs it is possible to find effective and efficient EAs. However, there is no general recipe for how to handle a CSP by EAs. Our experiments with graph coloring seem to confirm the conclusions of Richardson *et al* (1989) and Michalewicz (1996) that on tough problems the simple penalty function approach is not the best option.

An open research topic is whether one could forge EAs which construct a solution for a CSP, instead of searching in the space of complete instantiations of the variables. Actually, our application of decoders corresponds to deciding in which order the variables are instantiated (nodes are colored). Hence, the decoder technique can be seen as a method to construct different partial or complete instantiations of the variables. Whether it is possible to forge such EAs which operate on partial instantiations directly is an interesting research issue.

Optimal tuning of the EA (e.g. population size or mutation rates) remains an open issue. As the proper selection of these parameters very much depends on the characteristics of the solution space (how many solutions there are, how they are distributed), one can expect guidelines for specific types of CSP for which these characteristics have been investigated. On the basis of the characterization of the fitness landscape one may forecast the effectiveness and efficiency of a genetic operator (Manderick and Spiessens 1994).

We have given just one example of the possible benefits of adaptivity. Besides adapting penalties, EAs could also dynamically adjust parameters based on the evaluation of past experiences. Adaptively modifying operator probabilities (Davis 1989), mutation rates (Bäck 1992), or the population size (Arabas *et al* 1994) has led to interesting results. In addition to these parameters, similar techniques could be used to modify the heuristics, thus the genetic operators, based on earlier performance.

Finally, a hard nut is the question of unsolvable CSPs, since in general an EA cannot conclude for sure that a problem is not solvable. In the particular case of arc inconsistency the results of Bowen and Dozier (1995) are, however, very promising.

References

Arabas J, Michalewicz Z and Mulawka J 1994 GAVaPS—a genetic algorithm with varying population size *Proc. 1st. IEEE Int. Conf. on Evolutionary Computation (Orlando, FL, June 1994)* (Piscataway, NJ: IEEE) pp 306–11

Bäck T 1992 Self-adaptation in genetic algorithms *Proc. 1st European Conference on Artificial Life* ed F J Varela and P Bourgine (Cambridge, MA: MIT Press) pp 263–71

Bowen J and Dozier G 1995 Solving constraint-satisfaction problems using a genetic/systematic search hybrid that realizes when to quit *Proc. 6th Int. Conf. on Genetic Algorithms (Pittsburgh, PA, July 1995)* ed L J Eshelman (San Mateo, CA: Morgan Kaufmann) pp 122–9

Cheeseman P, Kenefsky B and Taylor W M 1991 Where the really hard problems are *Proc. IJCAI-91* (San Mateo, CA: Morgan Kaufmann) pp 331–7

Davis L 1989 Adapting operator probabilities in genetic algorithms *Proc. 3rd Int. Conf. on Genetic Algorithms (Fairfax, VA, June 1989)* ed J D Schaffer (San Mateo, CA: Morgan Kaufmann) pp 61–9

——1991 *Handbook of Genetic Algorithms* (New York: Van Nostrand Reinhold)

Dechter R 1990 Enhancement schemes for constraint processing: backjumping, learning, and cutset decomposition *Artificial Intell.* **41** 273–312

De Jong K A 1987 On using GAs to search problem spaces *Proc. 2nd Int. Conf. on Genetic Algorithms (Cambridge, MA, 1987)* ed J J Grefenstette (Hillsdale, NJ: Erlbaum) pp 210–6

Dozier G, Bowen J and Bahler D 1994 Solving small and large constraint-satisfaction problems using a heuristic-based microgenetic algorithms *Proc. 1st IEEE World Conf. on Evolutionary Computation (Orlando, FL)* (Piscataway, NJ: IEEE) pp 306–11

Eiben A E, Raue P-E and Ruttkay Zs 1994 Solving constraint-satisfaction problems using genetic algorithms *Proc. 1st IEEE World Conf. on Evolutionary Computation (Orlando, FL)* (Piscataway, NJ: IEEE) pp 542–7

——1995 Constrained problems *Practical Handbook of Genetic Algorithms* ed L Chambers (Boca Raton, FL: Chemical Rubber Company) pp 307–65

Eiben A E and Ruttkay Zs 1996 Self-adaptivity for constraint satisfaction: learning penalty functions *Proc. 3rd IEEE Conf. on Evolutionary Computation* (Piscataway, NJ: IEEE) pp 258–61

Eiben A E and van der Hauw J K 1996 *Graph Coloring with Adaptive Evolutionary Algorithms* Technical Report TR-96-11, Leiden University ftp://ftp.wi.leidenuniv.nl/pub/CS/TechnicalReports/1996/tr96-11.ps.gz

Fox B R and McMahon M B 1991 Genetic operators for sequencing problems *Proc. Workshop on the Foundations of Genetic Algorithms* ed G J E Rawlins (San Mateo, CA: Morgan Kaufmann) pp 284–300

Hao J K and Dorne R 1994 An empirical comparison of two evolutionary methods for satisfiability problems *Proc. 1st. IEEE Int. Conf. on Evolutionary Computation (Orlando, FL, June 1994)* (Piscataway, NJ: IEEE) pp 450–5

Mackworth A K 1977 Consistency in networks of relations *Artificial Intell.* **8** 99–118

Manderick B and Spiessens P 1994 How to select genetic operators for combinatorial optimization problems by analyzing their fitness landscapes *Computational Intelligence: Imitating Life* ed J M Zurada, R J Marks and C J Robinson (Piscataway, NJ: IEEE) pp 170–81

Meseguer P 1989 Constraint-satisfaction problems: an overview *AICOM* **2** 3–17

Michalewicz Z 1995 Genetic algorithms, numerical optimization and constraints *Proc. 6th Int. Conf. on Genetic Algorithms (Pittsburgh, PA, July 1995)* ed L J Eshelman (San Mateo, CA: Morgan Kaufmann) pp 151–8

——1996 *Genetic Algorithms + Data Structures = Evolutionary Computation* 3rd edn (Berlin: Springer)

Michalewicz Z and Attia N 1995 Evolutionary optimization of constrained problems *Proc. 3rd Ann. Conf. on Evolutionary Programming (San Diego, CA, February 1994)* ed A V Sebald and L J Fogel (Singapore: World Scientific) pp 98–108

Michalewicz Z and Michalewicz M 1995 Pro-life versus pro-choice strategies in evolutionary computation techniques *Computational Intelligence: a Dynamic System Perspective* ed M Palaniswami, Y Attikiouzel, R J Marks, D Fogel and T Fukuda (Piscataway, NJ: IEEE) pp 137–51

Minton S, Johnston M D, Philips A and Laird P 1992 Minimizing conflicts: a heuristic repair method for constraint satisfaction and scheduling problems *Artificial Intell.* **58** 161–205

Nudel B 1983 Consistent-labeling problems and their algorithms: expected complexities and theory based heuristics *Artificial Intell.* **21** 135–78

Olivier I M, Smith D J and Holland J C R 1987 A study of permutation crossover operators on the travelling salesman problem *Proc. 2nd Int. Conf. on Genetic Algorithms (Cambridge, MA, 1987)* ed J J Grefenstette (Hillsdale, NJ: Erlbaum) pp 224–30

Paredis J 1994 Co-evolutionary constraint satisfaction *Proc. 3rd Parallel Problem Solving from Nature* ed Y Davisor, H-P Schwefel and R Männer (Springer) pp 46–55

Richardson J T, Palmer M R, Liepins G and Hilliard M 1989 Some guidelines for genetic algorithms with penalty functions *Proc. 3rd Int. Conf. on Genetic Algorithms (Fairfax, VA, June 1989)* ed J D Schaffer (San Mateo, CA: Morgan Kaufmann) pp 191–7

Smith A E and Tate D M 1993 Genetic optimization using a penalty function *Proc. 5th Int. Conf. on Genetic Algorithms (Urbana-Champaign, IL, July 1993)* ed S Forrest (San Mateo, CA: Morgan Kaufmann) pp 499–505

Starkweather T, McDaniel S, Mathias K, Whitley D and Whitley Shaner C 1991 A comparison of genetic sequencing operators *Proc. 4th Int. Conf. on Genetic Algorithms (San Diego, CA, July 1991)* ed R K Belew and L B Booker (San Mateo, CA: Morgan Kaufmann) pp 69–76

Tsang E 1993 *Foundations of Constraint Satisfaction* (New York: Academic)

13

Niching methods

Samir W Mahfoud

13.1 Introduction

Niching methods (Mahfoud 1995a) extend genetic algorithms (GAs) to domains that require the location and maintenance of multiple solutions. While traditional GAs primarily perform optimization, GAs that incorporate niching methods are more adept at problems in classification and machine learning, multimodal function optimization, multiobjective function optimization (Chapter 5), and simulation of complex and adaptive systems.

Niching methods can be divided into families or categories, based upon structure and behavior. To date, two of the most successful categories of niching methods are *fitness sharing* (also called *sharing*) and *crowding*. Both categories contain methods that are capable of locating and maintaining multiple solutions within a population, whether those solutions have identical or differing fitnesses.

13.2 Fitness sharing

Fitness sharing, as introduced by Goldberg and Richardson (1987), is a fitness scaling mechanism that alters only the fitness assignment stage of a GA. Sharing can be used in combination with other scaling mechanisms, but should be the last one applied, just prior to selection.

From a multimodal function maximization perspective, the idea behind sharing is as follows. If similar individuals are required to share payoff or fitness, then the number of individuals that can reside in any one portion of the fitness landscape is limited by the fitness of that portion of the landscape. Sharing results in individuals being allocated to optimal regions of the fitness landscape. The number of individuals residing near any peak will theoretically be proportional to the height of that peak.

Sharing works by derating each population element's fitness by an amount related to the number of similar individuals in the population. Specifically, an element's *shared fitness*, F', is equal to its prior fitness F divided by its *niche count*. An individual's niche count is the sum of *sharing function* (sh) values between itself and each individual in the population (including itself). The shared fitness of a population element i is given by the following equation:

$$F'(i) = \frac{F(i)}{\sum_{j=1}^{\mu} \text{sh}(d(i, j))}.$$ (13.1)

The sharing function is a function of the distance d between two population elements; it returns a '1' if the elements are identical, a '0' if they cross some threshold of dissimilarity, and an intermediate value for intermediate levels of dissimilarity. The threshold of dissimilarity is specified by a constant, σ_{share}; if the distance between two population elements is greater than or equal to σ_{share}, they do not affect each other's shared fitness. A common sharing function is

$$\text{sh}(d) = \begin{cases} 1 - (d/\sigma_{\text{share}})^{\alpha} & \text{if } d < \sigma_{\text{share}} \\ 0 & \text{otherwise} \end{cases}$$ (13.2)

where α is a constant that regulates the shape of the sharing function.

While nature distinguishes its niches based upon phenotype, niching GAs can employ either *genotypic* or *phenotypic* distance measures. The appropriate choice depends upon the problem being solved.

13.2.1 Genotypic sharing

In genotypic sharing, the distance function d is simply the Hamming distance between two strings. (The Hamming distance is the number of bits that do *not* match when comparing two strings.) Genotypic sharing is generally employed by default, as a last resort, when no phenotype is available to the user.

13.2.2 Phenotypic sharing

In phenotypic sharing, the distance function d is defined using problem-specific knowledge of the phenotype. Given a function optimization problem containing k variables, the most common choice for a phenotypic distance function is Euclidean distance. Given a classification problem, the phenotypic distance between two classification rules can be defined based upon the examples to which they both apply.

13.2.3 Parameters and extensions

Typically, α is set to unity, and σ_{share} is set to a value small enough to allow discrimination between desired peaks. For instance, given a one-dimensional function containing two peaks that are two units apart, a σ_{share} of 1 is ideal: since each peak extends its reach for $\sigma_{\text{share}} = 1$ unit in each direction, the reaches of the peaks will touch but not overlap. Deb (1989) gives more details for setting σ_{share}.

Population size (Chapter 17)can be set roughly as a multiple of the number of peaks the user wishes to locate (Mahfoud 1995a, b). Sharing is best run for few generations, perhaps some multiple of $\log \mu$. This rough heuristic comes from shortening the expected convergence time for a GA that uses fitness-proportionate selection (Goldberg and Deb 1991). A GA under sharing will not converge population elements atop the peaks it locates. One way of obtaining such convergence is to run a hillclimbing algorithm after the GA.

Sharing can be implemented using any selection method, but the choice of method may either increase or decrease the stability of the algorithm. Fitness-proportionate selection with stochastic universal sampling (Baker 1987) is one of the more stable options. Tournament selection is another possibility, but special provisions must be made to promote stability. Oei *et al* (1991) propose a technique for combining sharing with binary tournament selection. This technique, *tournament selection with continuously updated sharing*, calculates shared fitnesses with respect to the new population as it is being filled.

The main drawback to using sharing is the additional time required to cycle through the population to compute shared fitnesses. Several authors have suggested calculating shared fitnesses from fixed-sized samples of the population (Goldberg and Richardson 1987, Oei *et al* 1991). Clustering is another potential remedy. Yin and Germay (1993) propose that a clustering algorithm be implemented prior to sharing, in order to divide the population into niches. Each individual subsequently shares only with the individuals in its niche. As far as GA time complexity is concerned, in real-world problems, a function evaluation requires much more time than a comparison; most GAs perform only $O(\mu)$ function evaluations each generation.

13.3 Crowding

Crowding techniques (De Jong 1975) insert new elements into the population by replacing similar elements. To determine similarity, crowding methods, like sharing methods, utilize a distance measure, either genotypic or phenotypic. Crowding methods tend to spread individuals among the most prominent peaks of the search space. Unlike sharing methods, crowding methods do not allocate elements proportional to peak fitness. Instead, the number of individuals congregating about a peak is largely determined by the size of that peak's basin of attraction under crossover.

By replacing similar elements, crowding methods strive to maintain the preexisting diversity of a population. However, replacement errors may prevent some crowding methods from maintaining individuals in the vicinity of desired peaks. The *deterministic crowding* algorithm (Mahfoud 1992, 1995a) is designed to minimize the number of replacement errors, and thus allow effective niching.

Deterministic crowding works as follows. First it groups all population elements into $\mu/2$ pairs. Then it crosses all pairs and mutates the offspring. Each offspring competes against one of the parents that produced it. For each pair of offspring, two sets of parent–child tournaments are possible. Deterministic crowding holds the set of tournaments that forces the most similar elements to compete.

The pseudocode for deterministic crowding is as follows:

Input: g—number of generations to run, μ—population size
Output: $P(g)$—the final population

```
P(0) ← initialize()
for t ← 1 to g do
    P(t) ← shuffle(P(t − 1))
    for i ← 0 to μ/2 − 1 do
        p₁ ← a_{2i+1}(t)
        p₂ ← a_{2i+2}(t)
        {c₁, c₂} ← recombine(p₁, p₂)
        c′₁ ← mutate(c₁)
        c′₂ ← mutate(c₂)
        if [d(p₁, c′₁) + d(p₂, c′₂)] ≤ [d(p₁, c′₂) + d(p₂, c′₁)] then
            if F(c′₁) > F(p₁) then a_{2i+1}(t) ← c′₁ fi
            if F(c′₂) > F(p₂) then a_{2i+2}(t) ← c′₂ fi
        else
            if F(c′₂) > F(p₁) then a_{2i+1}(t) ← c′₂ fi
            if F(c′₁) > F(p₂) then a_{2i+2}(t) ← c′₁ fi
        fi
    od
od
```

Deterministic crowding requires the user only to select a population size μ and a stopping criterion. As a general rule of thumb, the more final solutions a user desires, the higher μ should be. The user can stop a run after either a fixed number of generations g (of the same order as μ) or when the rate of improvement of the population approaches zero. Full crossover should be employed (with probability 1.0) since deterministic crowding only discards solutions after better ones become available, thus alleviating the problem of crossover disruption.

Two crowding methods similar in operation and behavior to deterministic crowding have been proposed (Cedeño *et al* 1994, Harik 1995). Cedeño *et al* suggest utilizing phenotypic crossover and mutation operators (i.e. *specialized* operators), in addition to phenotypic sharing; this results in further reduction of replacement error.

13.4 Theory

Much of the theory underlying sharing, crowding, and other niching methods is currently under development. However, a number of theoretical results exist, and a few areas of theoretical research have already been defined by previous authors. The characterization of hard problems is one area of theory. For niching methods, the number of optima the user wishes to locate, in conjunction with the number of optima present, largely determines the difficulty of a problem. A secondary factor is the degree to which extraneous optima lead away from desired optima.

Analyzing the distribution of solutions among optima for particular algorithms forms another area of theory. Other important areas of theory are calculating expected drift or disappearance times for desired solutions; population sizing; setting parameters such as operator probabilities and σ_{share} (for sharing); and improving the designs of niching genetic algorithms. For an extensive discussion of niching methods and their underlying theory, consult the article by Mahfoud (1995a).

References

Baker J E 1987 Reducing bias and inefficiency in the selection algorithm *Proc. Int. Conf. on Genetic Algorithms (Cambridge, MA)* ed J J Grefenstette (Hillsdale, NJ: Lawrence Erlbaum Associates) pp 14–21

Cedeño W, Vemuri V R and Slezak T 1994 Multiniche crowding in genetic algorithms and its application to the assembly of DNA restriction-fragments *Evolutionary Comput.* **2** 321–45

Deb K 1989 *Genetic Algorithms in Multimodal Function Optimization* Masters Thesis; TCGA Report 89002, University of Alabama, The Clearinghouse for Genetic Algorithms

De Jong K A 1975 *An Analysis of the Behavior of a Class of Genetic Adaptive Systems* Doctoral Dissertation, University of Michigan *Dissertation Abstracts Int.* **36** 5140B (University Microfilms 76-9381)

Goldberg D E and Deb K 1991 A comparative analysis of selection schemes used in genetic algorithms *Foundations of Genetic Algorithms* ed G J E Rawlins (San Mateo, CA: Morgan Kaufmann) pp 69–93

Goldberg D E and Richardson J 1987 Genetic algorithms with sharing for multimodal function optimization *Proc. 2nd Int. Conf. on Genetic Algorithms (Cambridge, MA)* ed J J Grefenstette (Hillsdale, NJ: Lawrence Erlbaum Associates) pp 41–9

Harik G 1995 Finding multimodal solutions using restricted tournament selection *Proc. 6th Int. Conf. on Genetic Algorithms (Pittsburgh, July 1995)* ed L J Eshelman (San Mateo, CA: Morgan Kaufmann) pp 24–31

Mahfoud S W 1992 Crowding and preselection revisited *Parallel Problem Solving From Nature* vol 2, ed R Männer and B Manderick (Amsterdam: Elsevier) pp 27–36

——1995a *Niching Methods for Genetic Algorithms* Doctoral Dissertation and IlliGAL Report 95001, University of Illinois at Urbana-Champaign, Illinois Genetic Algorithms Laboratory) *Dissertation Abstracts Int.* (University Microfilms 9543663)

——1995b Population size and genetic drift in fitness sharing *Foundations of Genetic Algorithms* vol 3, ed L D Whitley and M D Vose (San Francisco, CA: Morgan Kaufmann) pp 185–223

Oei C K, Goldberg D E and Chang S J 1991 *Tournament Selection, Niching, and the Preservation of Diversity* (IlliGAL Report 91011) University of Illinois, Illinois Genetic Algorithms Laboratory

Yin X and Germay N 1993 A fast genetic algorithm with sharing scheme using cluster analysis methods in multimodal function optimization *Artificial Neural Nets and Genetic Algorithms: Proc. Int. Conf. (Innsbruck)* ed R F Albrecht, C R Reeves and N C Steele (Berlin: Springer) pp 450–7

14

Speciation methods

Kalyanmoy Deb (14.1–14.4) *and William M Spears* (14.5, 4.6)

14.1 Introduction

Kalyanmoy Deb

Despite some controversy, most biologists agree that a species is a collection of individuals which resemble each other more closely than they resemble individuals of another species (Eldredge 1989). It is also clear that the reproductive process of the sexually reproducing organisms causes individuals to resemble their parents, thereby maintaining a phenotypic similarity among individuals of the community or the *species*. Thus, there is a strong correlation among the reproductively coherent individuals and a phenotypically similar cluster of individuals. Since in evolutionary algorithms a population of solutions is used, artificial species of phenotypically similar solutions can be formed and maintained in the population by restricting their mating to that with similar individuals. Before we outline how to form and maintain multiple species in a population, let us discuss why it could be necessary to form species in the applications of evolutionary algorithms.

In Chapter 13, we saw that multiple optimal solutions in a multimodal optimization problem can be found simultaneously by forming artificial niches (subpopulations) in the population. Each niche can be considered to represent a peak (in the spirit of maximization problems). To capture a number of peaks simultaneously and maintain them for many generations, a niching method is used. Niching helps to emphasize and maintain solutions around multiple optima. However, in niching, the main emphasis is devoted to distributing the population members across different peaks. Thus, the niching technique cannot quite focus its search on each peak and find the exact optimal solutions efficiently. This is because some of the search effort is wasted in the recombination of interpeak solutions, which, in turn, may produce some *lethal* solutions representing none of the peaks. A speciation method used in evolutionary computation (EC) studies, on the other hand, restricts mating to that among like solutions (likeness can be defined phenotypically or genotypically) and discourages mating among solutions of different peaks. If the likeness is defined properly, two parent solutions chosen for mating are likely to represent the same peak. Thus, when like individuals mate with each other, the created children solutions are also similar to the parent solutions and are likely to be members of the same peak. This way, the restriction of mating to that among like solutions may reduce the creation of lethal solutions (which represent none of the peaks). This may allow the search to concentrate on each peak and help find the best or near-best optimum solution efficiently. However, in order to apply the speciation technique properly, solutions representing each peak must first be found. Thus, the speciation technique cannot be used independently. In the presence of both niching and speciation, niching finds and maintains subpopulations of solutions around multiple optima and the speciation technique allows us to make an inherent parallel search in each optimum to find multiple optimal solutions simultaneously.

Among the evolutionary algorithms, a number of speciation methods have been suggested and implemented in genetic algorithms (GAs). Of the earlier works related to mating restriction in GAs, Hollstien's (1971) inbreeding scheme where mating was allowed between similar individuals in his simulation of animal husbandry problems, Booker's (1982) taxon–exemplar scheme for restrictive mating in his simulation of learning pattern classes, Holland's suggestion of a tag–template scheme (Goldberg 1989), Sannier and Goodman's (1987) restrictive mating in forming separate coherent groups in a population, Deb's (1989) phenotypic and genotypic mating restriction schemes, and Spears' (1994) and Perry's (1984) speciation using tag bits are a few studies. In the following, we discuss some of the above speciation methods in more detail.

14.2   Booker's taxon–exemplar scheme

Kalyanmoy Deb

Booker (1982) used taxons and exemplars in his learning algorithm to reduce the formation of lethal individuals. He defined a taxon as a string (constructed over the three-letter alphabet {0, 1, #}, with a # matching a 0 or a 1). The population is initialized with taxon strings. In his *restricted mating policy*, he wanted to restrict mating among similar taxon strings, which were identified by calculating a match score of the taxon strings with a given exemplar binary string. He allowed partial match scores depending on the matching of the taxon and the exemplar. For the following two taxon strings and the exemplar string, the first taxon matches the exemplar completely. The second taxon matches the exemplar partially (in first, third, and fourth positions):

$$\begin{array}{cc} \text{Taxon} & \text{Exemplar} \\ (1\ \#\ 0\ 0\ \#) & (1\ 0\ 0\ 0\ 0). \\ (\#\ 1\ \#\ 0\ 1) & \end{array}$$

If the taxon completely matches the exemplar, a score is assigned as the sum of the string length and the number of #s in the taxon. The partial credit is also assigned based on the number of correct matches and the number of #s in the taxon. In order to implement the restrictive mating policy, he chose parent taxon strings from a sample subpopulation determined based on the available matching taxon strings in the population. If a specified number of matching taxon strings are available in the population, parent strings are chosen uniformly at random from all the matching taxon strings. Otherwise, parent strings are chosen according to a probability distribution calculated based on the match score of the taxon strings. In a number of pattern discovery problems an improved performance is observed with the restricted mating policy.

After the patterns were discovered, Booker extended his above scheme to classify the discovered patterns using a modified string as follows:

$$\begin{array}{cc} \text{Taxon} & \text{Tag} \\ (1 \ \# \ 0 \ 0 \ \#) & : \quad (1 \ 0 \ 0 \ 0 \ 0). \end{array}$$

In addition to the taxon string, a tag string is introduced to classify the discovered taxon strings (or patterns). The taxon strings matching a particular tag string were considered to be in the same class. A similar match score was used, except that this time the matching was performed with the taxon and tag strings. As discussed elsewhere (Goldberg 1989), there is one difficulty with the above tag–taxon scheme. The tag string must be of the same length as the taxon string. This increases the complexity of the classification problem, whereas the same concept can be implemented with shorter tag and template strings, as suggested by Holland; a brief description of this is given by Goldberg (1989).

14.3   The tag–template method

Kalyanmoy Deb

In addition to the functional string (the taxon string in Booker's pattern classification problem), a template and a tag string are introduced. The template string is constructed from the three-letter alphabet (1, 0, and #) as before, but the tag string is a binary string of the same length as the template string. A typical string with the tag and template strings would look like the following:

$$\begin{array}{ccc} \text{Template} & \text{Tag} & \text{Functional string} \\ (\#01) & : \quad (100) & : \quad (1011001101). \end{array}$$

The size of tag and template strings depends on the number of desired solutions. A simple calculation shows that if q different optimal solutions (peaks) are to be found, the minimum string length for the tag and template is $\lceil \log_2 q \rceil$ (Deb 1989). The tag and template strings are created at random in the initial population along with the functional string. These two strings do not affect the fitness of the functional string. However, they are affected by the crossover and the mutation operators, as well. For the template string, the mutation operator must be modified to operate on a three-allele string. The purpose of these strings is to restrict mating. Before crossing a pair of individual strings, their tag and template strings are *matched*. If the match score exceeds a threshold value, the crossover is performed between the two strings as usual; otherwise some other string pair is tested for a possible mating. In this process, the tag and template strings corresponding to the good individuals in early populations are emphasized and an artificial tag is set for solutions in each peak. Later on, since crossing over is only performed between the matched strings, only similar strings (or strings from the same peak) tend to participate in crossover.

Although neither Holland nor Goldberg simulated this speciation method, Deb (1989) (with assistance from David Goldberg) implemented this scheme and applied this technique in solving multimodal test problems. In both cases, GAs with the tag–template scheme performed better than GAs without it.

14.4   Phenotypic and genotypic mating restriction

Kalyanmoy Deb

Deb (1989) has developed two mating restriction schemes based on the phenotypic and genotypic distance between mating individuals. The mating restriction schemes are straightforward. In order to choose a mate for an individual, their distance (in phenotypic mating restriction the Euclidean distance and in genotypic mating restriction the Hamming distance) is computed. If the distance is closer than a parameter σ_{mating}, they participate in the crossover operation; otherwise another individual is chosen at random and their distance is computed. This process is continued until a suitable mate is found or all population members are exhausted, in which case a random individual is chosen as a mate. Deb has implemented both the above mating restriction schemes with a single-point crossover and applied them to solve a number of multimodal test problems. Although, in all his simulations, the parameter σ_{mating} was kept the same as the parameter σ_{share} used in the niching methods, other values of σ_{mating} may also be chosen. It is worthwhile to mention that niching with the σ_{share} parameter is implemented in the selection operator and the mating restriction with the σ_{mating} parameter is implemented in the crossover operator. GAs with niching and mating restriction were found to better distribute the population across the peaks than GAs with sharing alone. Here, we present simulation results for the phenotypic mating restriction scheme adopted in that study. In solving the single-variable, five-peaked function in the interval $0 \leq x \leq 1$

$$\text{maximize} \quad 2^{-2((x-0.1)/0.8)^2} \sin^6(5\pi x)$$

with $\sigma_{share} = \sigma_{mating} = 0.1$, 100 population members after 200 generations without and with phenotypic mating restriction are shown in figure 14.1. Stochastic remainder roulette wheel selection and single-point crossover

operators are used.   The crossover and mutation probabilities are kept as
0.9 and 0.0, respectively. The figures show that, with the mating restriction
scheme (the right-hand panel), the number of lethal (nonpeak) individuals
has been significantly decreased.   This study also implemented a genotypic
mating restriction scheme and similar results were obtained. Some guidelines in
choosing the sharing and mating restriction parameters are outlined elsewhere
(Deb 1989, Deb and Goldberg 1989).

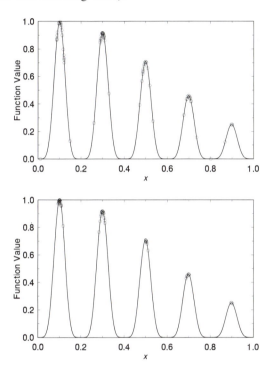

Figure 14.1. The distribution of 100 solutions without (left) and with (right) a mating
restriction scheme.

14.5   Speciation using tag bits

William M Spears

Another method for identifying species is via the use of tag bits, which are
appended to every individual. Each species corresponds to a particular setting
of these bits. Suppose there are k different sets of tag bit values at a particular
generation of the evolutionary algorithm (EA). Denote these sets as $\{S_0, \ldots,
S_{k-1}\}$. The sets are numbered arbitrarily. Each individual belongs to one S_i
and all individuals in a particular S_i have the same tag bit values. For example,

suppose there is only one tag bit and that some individuals exist with a tag bit value zero and that the remainder exist with tag bit value one. Then (arbitrarily) assign the former set of individuals to S_0 and the latter set to S_1. Let $\| \ \|$ denote the cardinality of the sets.

Spears (1994) uses the tag bits to restrict mating and to perform fitness sharing. With sharing, the perceived fitness, F_i, is a normalization of the objective fitness f_i:

$$F_i = \frac{f_i}{\|S_j\|} \qquad i \in S_j$$

where $\|S_j\|$ is the size of the species that individual i is in.

The average fitness of the population, $\bar{F}$, becomes

$$\bar{F} = \frac{\sum_{i \in S_0} (f_i / \|S_0\|) + \ldots + \sum_{i \in S_{k-1}} (f_i / \|S_{k-1}\|)}{\|S_0\| + \ldots + \|S_{k-1}\|}$$

which is just

$$\bar{F} = \frac{\sum_{i \in S_0} (f_i / \|S_0\|) + \ldots + \sum_{i \in S_{k-1}} (f_i / \|S_{k-1}\|)}{N}$$

since the species sizes have to total N (recall that no individual can lie in more than one species). The expected number of offspring for an individual is now $F_i / \bar{F}$.

Restricted mating is performed by only allowing recombination to occur between individuals with the same tag bit values. Mutation can flip all bits, including the tag bits, thus allowing individuals to change labels. Experimental results, as well as some modifications to the above mechanism can be found in the article by Spears (1994). Code for the algorithm can be found at http://www.aic.nrl.navy.mil/~spears.

Perry's thesis work (Perry 1984) with speciation is extremely similar to the above technique. Perry includes both species and environmental regions in an EA. Species are identified via tag bits and an environmental region is similar to an EA population. Recombination within an environment can occur only on individuals with the same tag bit values. Mutation is allowed to change tag bits, in order to introduce new species. The additional use of a 'migration' operator, which moves individuals from one environment to another, does not have an analog in the work of Spears (1994).

Perry gives an example of two species in an environment—fitness proportional selection is performed, and the average fitness of an environment is

$$\bar{f} = \frac{\sum_{i \in S_0} f_i + \sum_{i \in S_1} f_i}{\|S_0\| + \|S_1\|}$$

or

$$\bar{f} = \frac{\sum_{i \in S_0} f_i + \sum_{i \in S_1} f_i}{N}$$

where N is the population size of the environmental niche. The expected number of offspring is $f_i / \bar{f}$. One can see that the main difference between the two methods is the use of sharing in the computation of fitness in the work of Spears (1994). Thus it is not surprising that in many of Perry's experimental runs one particular species would eventually dominate an environmental niche (however, it should be noted that in the work of Perry (1984) the domination of an environment by a species was not undesirable behavior).

The use of tag bits makes restricted mating and fitness sharing more efficient because distance comparisons do not have to be computed. Interestingly, it is also possible to make Goldberg's implementation of sharing more efficient by sampling (Goldberg *et al* 1992). In other words the distance of each individual from the rest is estimated by using a subset of the remaining individuals.

14.6 Relationship with parallel algorithms

William M Spears

Clearly this work has similarities to the EA research performed on parallel architectures (Section 15.1, Chapter 16). In a parallel EA, a topology is imposed on the EA population, resulting in species. However, there are some important differences between the parallel approaches and the sequential approach. For example, with the fitness sharing approaches the fitness of an individual and the species size are dynamic, based on the other individuals (and species). This concentrates effort on more promising peaks, while still maintaining individuals in other areas of the search space. This is typically not true for parallel EAs implemented on MIMD or SIMD architectures. When using a MIMD architecture, species are dedicated to particular processors and the species remain a constant size. In SIMD implementations, one or two individuals reside on a processor, and species are formed by defining overlapping neighborhoods. However, due to the overlap, one particular species will eventually take over the whole population.

References

Booker L B 1982 *Intelligent Behavior as an Adaptation to the Task Environment* Doctoral Dissertation, University of Michigan; *Dissertation Abstracts Int.* **43** 469B

Deb K 1989 *Genetic Algorithms in Multimodal Function Optimization* Master's Thesis, University of Alabama; TCGA Report 89002

Deb K and Goldberg D E 1989 An investigation of niche and species formation in genetic function optimization *Proc. 3rd Int. Conf. on Genetic Algorithms (Fairfax, VA, 1989)* ed J D Schaffer (San Mateo, CA: Morgan Kaufmann) pp 42–50

Eldredge N 1989 *Macro-evolutionary Dynamics: Species, Niches and Adaptive Peaks* (New York: McGraw-Hill)

Goldberg D E 1989 *Genetic Algorithms in Search, Optimization, and Machine Learning* (Reading, MA: Addison-Wesley)

Goldberg D E and Richardson J 1987 Genetic algorithms with sharing for multimodal function optimization *Proc. 2nd Int. Conf. on Genetic Algorithms (Cambridge, MA, 1987)* ed J J Grefenstette (Hillsdale, NJ: Erlbaum) pp 41–9

Goldberg D E, Deb K and Horn J 1992 Massive multimodality, deception, and genetic algorithms *Proc. Parallel Problem Solving from Nature Conf.* (Amsterdam: North-Holland) pp 37–46

Hollstien R B 1971 *Artificial Genetic Adaptation in Computer Control Systems* Doctoral Dissertation, University of Michigan; *Dissertation Abstracts Int.* **32** 1510B

Perry Z A 1984 *Experimental Study of Speciation in Ecological Niche Theory using Genetic Algorithms* Doctoral Dissertation, University of Michigan; *Dissertation Abstracts Int.* **45** 3870B

Sannier A V and Goodman E D 1987 Genetic learning procedures in distributed environments *Proc. 2nd Int. Conf. on Genetic Algorithms (Cambridge, MA, 1987)* ed J J Grefenstette (Hillsdale, NJ: Erlbaum) pp 162–9

Spears W M 1994 Simple subpopulation schemes *Proc. Conf. on Evolutionary Programming* (Singapore: World Scientific) pp 296–307

15

Island (migration) models: evolutionary algorithms based on punctuated equilibria

W N Martin, Jens Lienig and James P Cohoon

15.1 Parallelization

Research to develop parallel implementations (Chapter 24) of algorithms has a long history (Slotnick *et al* 1962, Barnes *et al* 1968, Wulf and Bell 1972) across many disparate application areas. The majority of this research has been motivated by the desire to reduce the overall time to completion of a task by distributing the work implied by a given algorithm to processing elements working in parallel. More recently some researchers have conjectured that some parallelizations of a task improve the quality of solution obtained for a given overall amount of work, e.g. emergent computation (Forrest 1991), and some even suggest that considering parallelization may lead to fundamentally new modes of thought (Bailey 1992). Note that the benefits of this latter kind of parallelization depend only on concurrency, i.e. the logical temporal independence, of operations and thus they can also be obtained via sequential simulations of parallel formulations.

The more prevalent motivation for parallelization, i.e. reducing time to completion, depends on the specifics of the architecture executing the parallelized algorithm. Very early on, it was recognized that different parallel hardware made possible different categories of parallelization based on the *granularity* of the operations performed in parallel. Typically these categories are referred to as *fine-grained*, *medium-grained*, and *coarse-grained* parallelization. At the extremes of this spectrum, fine-grained (or small-grained) parallelism means that only short computation sequences are performed between synchronizations, while coarse-grained (or large-grained) parallelism means that extended computation sequences are performed between synchronizations. SIMD (single-instruction, multiple-data) architectures are most appropriate for fine-grained parallelism (Fung 1976), while distributed-memory message-passing architectures are most appropriate for coarse-grained parallelism (Seitz 1985).

Two of the earliest parallelizations of a genetic algorithm (GA) were based on a distributed-memory message-passing architecture (Tanese 1987, Pettey *et al* 1987; also see Grosso (1985) for an early serial simulation of a concurrent formulation). The resulting parallelization was coarse grained in that the overall population of the GA was broken into a relatively small number of subpopulations. Each processing element in the architecture was assigned an entire subpopulation and executed a rather standard GA on its subpopulation.

In the same time frame, it was noted that a theory concerning speciation and stasis in populations of living organisms, called *punctuated equilibria*, provided evidence that in natural systems this kind of *parallelization* of evolution had an emergent property of bursts of rapid evolutionary *progress*. The resulting parallel GA was shown to have this property on several applications (Cohoon *et al* 1987).

All of the above systems are examples of what has come to be called *island model* parallel genetic algorithms (Gordon *et al* 1992, Adamidis 1994). In the next section we discuss theories of natural evolution as they support and motivate island model formulations. We then discuss the important aspects, parameters, and attributes of systems built on this model. Finally, we present results of one such system on a difficult very large-scale integration (VLSI) design problem.

15.2   Theories of natural evolution

In what has been called the *modern synthesis* (Huxley 1942), the fields of biological evolution and genetics began to be merged. A major development in this synthesis was Sewall Wright's (1932) conceptualization of the adaptive landscape. The original conceptualization proposes an underlying space (two-dimensional for discussion purposes) of possible genetic combinations. At each point in that space an *adaptive value* is determined and specified as a scalar quantity. The surface thus specified is referred to as the *adaptive landscape*. A population of organisms can be mapped to the landscape by taking each member of the population, determining the point in the underlying space that its genetic code specifies, and marking the associated surface point. The figure used repeatedly by Wright shows the adaptive landscape as a standard topographic map with contour lines of equal adaptive value instead of altitude. The + symbols indicate local maxima. A population—in two demes—is then depicted by two shaded regions overlaid on the map.

There are several reasons why we used the word *conceptualization* in the previous paragraph. First and foremost, it is not clear what the topology of the underlying space should be. Wright (1932) considers initially the individual gene sequences and connects genetic codes that are 'one remove' from each other, implying that the space is actually an undirected connected graph. He then turns immediately to a continuous space with each gene locus specifying a dimension and with units along each dimension being the possible allelomorphs at the given locus. Specifying the underlying space to be an multidimensional

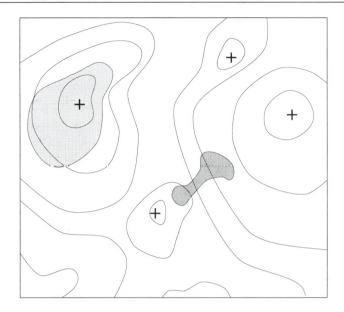

Figure 15.1. An adaptive landscape according to Wright, with a sample population—in two demes (shaded areas).

Euclidean space determines the topology. However, if one is to attempt to make inferences from the character of the adaptive landscape (Radcliffe 1991), the ordering of the units along the various dimensions is crucial. With arbitrary orderings the metric notions of *nearby* and *distant* have no clear-cut meaning; similar ambiguities occur in many discrete optimization problems. For instance, given two tours in a traveling salesperson problem, what is the proper measure of their closeness?

The concept of the adaptive landscape has had a powerful effect on both microevolutionary and macroevolutionary theory, as well as providing a fundamental basis for considering genetic algorithms as function optimizers. As Wright states (1932):

> The problem of evolution as I see it is that of a mechanism by which the species may continually find its way from lower to higher peaks in such a field. In order that this may occur, there must be some trial and error mechanism on a grand scale ...

Wright also used the adaptive landscape concept to explain his mechanism, the *shifting balance theory*. In the shifting balance theory the ability for a species to 'search' and not be forced to remain at lower adaptive peaks by strong selection pressure is provided through a population structure that allows the species to take advantage of ecological opportunities. The population structure is based upon *demes*, as Wright describes (1964):

> Most species contain numerous small, random breeding local
> populations (demes) that are sufficiently isolated (if only by distance)
> to permit differentiation ...

Wright conceives the shifting balance to be a microevolutionary mechanism,
that is, a mechanism for evolution within a species. For him the emergence of
a new species is a corollary to the general operation and progress of the shifting
balance. Eldredge and Gould (1972) have contended that macroevolutionary
mechanisms are important and see the emergence of a new species as being
associated very often with extremely rapid evolutionary development of diverse
organisms. As Eldredge states (1989):

> Other authors have gone further, suggesting that SMRS [SMRS
> denotes *specific mate recognition system*, the disruption of which is
> presumed to cause reproductive isolation] disruption actually may
> *induce* [his emphasis] economic adaptive change, i.e., rather than
> merely occur in concert with it, ... [Eldredge and Gould] have
> argued that small populations near the periphery of the range of an
> ancestral population may be ideally suited to rapid adaptive change
> following the onset of reproductive isolation ... Thus SMRS disruption
> under such conditions may readily be imagined to act as a 'release,'
> or 'trigger' to further adaptive change the better to fit the particular
> ecological conditions at the periphery of the parental species's range.

The island model GA formulation by Cohoon *et al* (1987, 1991a) was strongly
influenced by this theory of punctuated equilibria (Eldredge and Gould 1972),
so they dubbed the developed system the *genetic algorithm with punctuated
equilibria* (GAPE). In general, the important aspect of the Eldredge–Gould
theory is that one should look to small disjoint populations, i.e. peripheral
isolates, for extremely rapid evolutionary change.

For the analogy to discrete optimization problems, the peripheral isolates
are the semiindependent subpopulations and the rapid evolutionary change is
indicative of extensive search of the solution domain. Thus, we contend
that the island model genetic algorithm is rightly considered to be based
on a population structure that involves subpopulations which have their
isolated evolution occasionally punctuated by interpopulation communication
(Cohoon *et al* 1991b). To relate these processes to Holland's terms (1975),
the *exploration* needed in GAs arises from the infusion of migrants, i.e.
individuals from neighboring subpopulations, and the *exploitation* arises from
the isolated evolution. It is this alternation between phases of communication
and computation that holds the promise for island model GAs to be more than
just hardware accelerators for the evolutionary process. In the next section the
major aspects of such island models will be delineated.

15.3   The island model

The basic model begins with the islands—the *demes* (Wright 1964) or the *peripheral isolates* (Eldredge and Gould 1972). Here the islands will be referred to as *subpopulations*. It is important to note again that while one motivation in parallelization would demand that each subpopulation be assigned to its own processing element, the islands are really a logical structure and can be implemented efficiently on many different architectures. For this reason we will refer to each subpopulation being assigned to a process, leaving open the issue of how that process is executed.

The island model will consider there to be an overall population $\mathcal{P}$ of $M = |\mathcal{P}|$ individuals that is partitioned into N subpopulations $\{\mathcal{P}_1, \mathcal{P}_2, \ldots, \mathcal{P}_N\}$. For an even partition each subpopulation has $\mu = M/N$ individuals, but for generality we can have $|\mathcal{P}_i| = \mu_i$ so that each subpopulation might have a distinct size. For standard GAs the selection of M can be problematic and for island model GAs this decision is compounded by the necessity to select N (and thereby μ). In practice, the decisions often need to be made in the opposite order; that is, μ_i is crucial to the dynamics of the trajectory of evolution for $\mathcal{P}_i$ and is heavily problem dependent. We believe that for specific problems there is a threshold size, below which poor results are obtained (as we will show in Section 15.5). Further, we believe that island model GAs are less sensitive to the choice of μ_i, as long as it is above the threshold for the problem instance. With μ_i decided, the selection of N (and thereby M) is often based on the available parallel architecture.

Given N, the next decision is the subpopulation interconnection. This is generally referred to as the *communication topology*, in that the island model presumes migration, i.e. intersubpopulation communication. The $\mathcal{P}_i$ are considered to be the vertices of a graph (usually undirected or at least symmetric) with each edge specifying a communication link between the incident vertices. These links are often taken to correspond to actual communication links between the processing elements assigned to the subpopulations. In any case, the communication topology is almost always considered to be static.

Given the ability for two subpopulation processes to communicate, the magnitude and frequency of that communication must be determined. Note that if one allows zero to be a possible magnitude then the communication topology and magnitudes can be specified by a matrix S, where S_{ij} is the number of individuals sent from $\mathcal{P}_i$ to $\mathcal{P}_j$. $S_{ij} = 0$ indicates no communication edge.

As was mentioned in Section 15.2, the migration pattern is important to the overall evolutionary trajectory. The migration pattern is determined by the degree of connectivity in the communication topology, the magnitude of communication, and the frequency of communication. These parameters determine the amount of isolation and interaction among the subpopulations. The parameters are important with regard to both the *shifting balance* and *punctuated equilibria* theories. Note that as the connectivity of the topology increases, i.e. tends toward a completely connected graph, and the frequency of interaction increases, i.e. the isolated evolution time for each $\mathcal{P}_i$ is shortened, the island model approximates more closely a single, large, freely intermixing population (see Section 15.5). It is held generally that such large populations quickly reach stable gene frequencies, and thus, cease 'progress'. Eldredge and Gould (1972) termed this *stasis*, while the GA community generally refers to it as *premature convergence*.

At the other extreme, as the connectivity of the topology decreases, i.e. tends toward an edgeless graph, and the frequency of interaction decreases, i.e. each $\mathcal{P}_i$ has extended isolated evolution, the island model approximates more closely several independent trials of a sequential GA with a small population. We contend that such small populations 'exploit' strongly the area of local optima, but only those local optima extremely 'close' to the original population. Thus, intermediate degrees of connectivity and frequency of interaction provide the dynamics sufficient to allow both exploitation and exploration.

For our discussion here, the periods of isolated evolution will be called *epochs*, with migration occurring at the end of each epoch (except the last). The length of the epochs determines the frequency of interaction. Often the epoch length is specified by a number G_i of generations that $\mathcal{P}_i$ will evolve in isolation. However, a formulation more faithful to the theories of natural evolution would be to allow each subpopulation process to reach stasis, i.e. reach equilibrium or convergence, on each epoch (see Section 15.5). From an implementation point of view with a subpopulation assigned to each processing element, this latter formulation allows the workload to become unbalanced and as such may be seen as an inefficient use of the parallel hardware if the processing elements having quickly converging subpopulations are forced to sit idle. The more troublesome problem is in measuring effectively the degree of stasis. 'Inefficiency' might occur when reasonably frequent, yet consistently marginal 'progress' is being made. Then not only might other processing elements be idle, but also the accumulated progress might not be worth the computation spent. In one of the experiments of the next section, we will present a system that incorporates an epoch-termination criterion. This system yields high-quality results more consistently, while being implemented in an overall parallel computing environment that utilizes the 'idle' processing elements.

The overall structure of the island model process comprises E major iterations called epochs. During an epoch each subpopulation process independently executes a sequential evolutionary algorithm for G_i generations. After each epoch there is a communication phase during which individuals migrate between neighboring subpopulations. This structure is summarized in the following pseudocode:

Island_Model(E,N,μ)

{

   Concurrently for each of the $i \leftarrow 1$ to N subpopulations
   Initialize($\mathcal{P}_i, \mu$);
   For *epoch* $\leftarrow 1$ to E do
         Concurrently for each of the $i \leftarrow 1$ to N subpopulations do
                         Sequential_EA($\mathcal{P}_i, G_i$);
         od;
         For $i \leftarrow 1$ to N do
                         For each neighbor j of i
                                 Migration($\mathcal{P}_i, \mathcal{P}_j$);
                         Assimilate($\mathcal{P}_i$);
         od
   od
   problem solution = best individual of all subpopulations;

}

Note that we specified **Sequential_EA** because the general framework can be applied to other evolutionary algorithms, e.g. evolution strategies (ES) (Lohmann 1990, Rudolph 1990).

After each phase of migration each subpopulation must assimilate the migrants. This assimilation step is dependent on the details of the migration process. For instance, in the implemented island model presented in the next section, if individual p_k is selected for emigration from $\mathcal{P}_i$ to $\mathcal{P}_j$ then p_k is deleted from $\mathcal{P}_i$ and added to $\mathcal{P}_j$. (The individual p_k itself migrates.) Also, the migration magnitudes, S_{ij}, are symmetric. Thus, the size of each subpopulation remains the same after migration and the assimilation is simply a fitness recalculation.

In other island models (Cohoon *et al* 1991a), if p_k is selected for emigration from $\mathcal{P}_i$ to $\mathcal{P}_j$ then p_k is added to $\mathcal{P}_j$ without being removed from $\mathcal{P}_i$. (A copy of individual p_k migrates.) This migration causes the subpopulation size to increase, so (under an assumption of a constant-size-subpopulation GA) the assimilation must include a reduction operation.

Still other parallel GAs (Mühlenbein *et al* 1987, Mühlenbein 1989, Gorges-Schleuter 1990, 1992, Tamaki 1992) implement overlapping subpopulations, i.e. the diffusion model (Chapter 16). For such systems migration is not really an issue; rather its important effect is attained through the selection process. However, parallel GAs with overlapping subpopulations are best suited to medium-grained parallel architectures.

An important aspect of both the *shifting balance* and *punctuated equilibria* theories is that the demes, or peripheral isolates, evolve in distinct environments. This aspect has two major facets. The first, and most obvious, facet is the restriction of the available breeding population, i.e. the isolated evolution of each subpopulation as we have already discussed. The second facet is the differing environmental attributes that determine the factors in natural selection. Wright has suggested that this facet provides 'ecological opportunity' (1982). For most GAs these factors, and the ways they interrelate to form the basis of natural selection, are encoded in the fitness function. Thus, to follow the fundamental analogy, the island model should have differing fitness functions at the various subpopulations. Of course, for GAs that are used for function optimization, the fitness function is almost always the objective function to be optimized (or some slight variation, e.g. an inversion to make an original minimization consistent with 'fitness'). We are not aware of any systems that have made use of this facet with truly distinct fitness functions among the subpopulations.

The island model presented in the next section (and many others) has a rudimentary form of this facet through *local* normalization of objective scores to yield fitness values. For example, an individual's fitness might be assigned to be its raw objective score divided by the current mean objective score across the given subpopulation. (This is the reason for the two fitness calculations in the pseudocode presented in subsection 15.4.2.) Such normalization does effect differing environments to the degree that the distributions of the individuals in the subpopulations differ.

For optimization problems that are multiobjective, there is usually a rather arbitrary linear weighting of the various objective dimensions to yield a scalar objective score (see e.g. (15.1)). This seems to provide a natural mechanism for having distinct objective functions, namely, distinct coefficient sets at each subpopulation. The difficulty of using this mechanism is that it clearly adds another level of evolution control parameters. Eldredge and Gould recognized this additional level when they discussed punctuated equilibria as a theory about the evolution of species, not a theory about the evolution of individuals within a species (Eldredge and Gould 1972). This is, indeed, an important facet of the island model; unfortunately, further exploration of its form and implications is beyond the scope of our discussion here.

15.4   The island model genetic algorithm applied to a VLSI design problem

In order to illustrate the effectiveness of this parallel method and the effects of modifying important island model parameters, we present an island model applied to the routing problem in VLSI circuit design. In Section 15.5, we then present the results from experiments in which the selected parameters were varied systematically and overall system performance evaluated.

15.4.1   Problem formulation

The VLSI routing problem is defined as follows. Consider a rectangular routing region on a VLSI circuit with *pins* located on two parallel boundaries (*channel*)

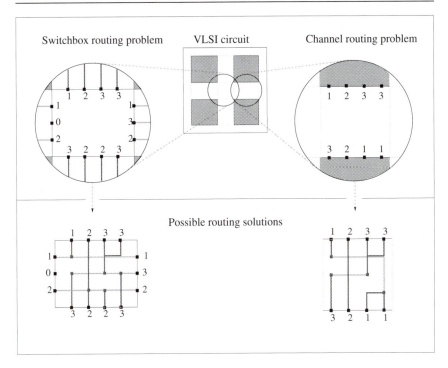

Figure 15.2. Example switchbox and channel routing problems ('magnified' in the circles) and possible routing solutions.

or four boundaries (*switchbox*). The pins that belong to the same *net* need to be connected subject to certain constraints and quality factors. The interconnections need to be made inside the boundaries of the routing region on a symbolic routing area consisting of horizontal *rows* and vertical *columns* (see figure 15.2).

The routing quality of a particular solution involves (for the purposes of the following experiments) three factors: *netlength*—the shorter the length of the interconnections, the smaller the propagation delay, *number of vias*—the smaller the number of vias (connections between routing layers), the fewer electrical and fabrication problems occur, and *crosstalk*—in submicrometer regimes, crosstalk results mainly from coupled capacitance between adjacent (parallel routed) interconnections, so the shorter these parallel-routed segments are, the less crosstalk occurs and the better the performance of the circuit. Thus, the optimization is to find a routing solution p_i for which $Obj(p_i)$ is minimal, with the objective function Obj specified by

$$Obj(p_i) = w_1 * l_{\text{nets}}(p_i) + w_2 * n_{\text{vias}}(p_i) + w_3 * l_{\text{par}}(p_i) \qquad (15.1)$$

where $l_{\text{nets}}(p_i)$ is the total length of the nets of p_i, $n_{\text{vias}}(p_i)$ is the number of vias of p_i, $l_{\text{par}}(p_i)$ is the total length of adjacent, parallel-routed net segments of p_i (crosstalk segments), and w_1, w_2, and w_3 are weight factors.

For VLSI designers it is important to have the weight factors, i.e. w_1, w_2, and w_3, to enable the designer to easily adjust routing quality characteristics: the netlength, the number of vias, and the tolerance of crosstalk, respectively, to the requirements of a given VLSI technology.

15.4.2   The evolutionary process

As indicated in the pseudocode in Section 15.3, each subpopulation process executes a sequential evolutionary algorithm. In this application we incorporated a genetic algorithm specified by the following pseudocode:

Sequential_GA($\mathcal{P}_i$, G_i)
{
   For *generation* $\leftarrow$ 1 to G_i do
         $\mathcal{P}_{\text{new}} \leftarrow \emptyset$;
         For *offspring* $\leftarrow$ 1 to *Max_offspring$_i$* do
             $p_\alpha \leftarrow$ **Selection**($\mathcal{P}_i$);
             $p_\beta \leftarrow$ **Selection**($\mathcal{P}_i$);
             $\mathcal{P}_{\text{new}} = \mathcal{P}_{\text{new}} \cup$ **Crossover**(p_α, p_β);
         od
         Fitness_calculation($\mathcal{P}_i \cup \mathcal{P}_{\text{new}}$);
         $\mathcal{P}_i \leftarrow$ **Reduction**($\mathcal{P}_i \cup \mathcal{P}_{\text{new}}$);
         Mutation($\mathcal{P}_i$);
         Fitness_calculation($\mathcal{P}_i$);
   od
}

Here $\mathcal{P}_i$ is the initial subpopulation that already includes any assimilated migrants from the last migration phase. In addition, the GA indicated above requires the following problem-domain-specific operators. For these operators, each individual is a complete routing solution, p_i.

Initialization.   A random routing strategy (Lienig and Thulasiraman 1994) is used to create the initial subpopulations consisting of 'nonoptimized' routing solutions. These initial routing solutions are guaranteed to be feasible solutions, i.e. all necessary connections exist, but no refinement is performed on them. Thus, we consider them to be random solutions that are distributed throughout the search space.

Fitness calculation.   The higher-quality solutions have smaller objective function values. So, to get a fitness value suitable for maximizing, a raw fitness function is calculated as the inverse of the objective function (see equation (15.1)), $F'(p_i) = 1/Obj(p_i)$, then the final fitness $F(p_i)$ of each individual

p_i is determined from $F'(p_i)$ by linear scaling (Goldberg 1989) *local* to the specific subpopulation.

Selection.   The selection strategy, which is responsible for choosing mates for the crossover procedure, is stochastic sampling with replacement (Goldberg 1989); that is, individuals are selected with probabilities proportional to their fitness values.

Crossover.   Two individuals are combined to create a single offspring. The crossover operator gives high-quality routing components of the parents an increased probability of being transferred intact to their offspring (low disruption). The operator is analogous to one-point crossover, with a randomly positioned line (a horizontal or vertical crossline) that divides the routing area into two sections, playing the role of the crosspoint. For example, net segments *exclusively* on the upper side of a horizontal crossline are inherited from the first parent, while segments *exclusively* on the lower side of the crossline are inherited from the second parent. Net segments intersecting the crossline are newly created for the offspring by means of a random routing strategy (the same strategy as used in *initialization*). The full details of this operator (Lienig and Thulasiraman 1994) are beyond the scope of this discussion, but figure 15.3 provides a general idea of how each individual routing solution is represented. In the genotype, the routing surface is represented by an 'occupancy' model, that is, each unit of surface area is represented by a node (a circle in figure 15.3). Nodes are connected if their corresponding surface areas are adjacent, either within a layer or across layers. The value at a node indicates which net is routed through that surface area, with a negative value indicating a pin (thus fixed assignment) position and zero indicating that the area is unused.

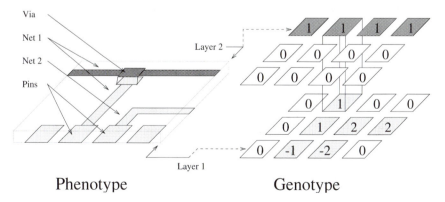

Figure 15.3. Representing a routing solution (the phenotype) as a three-dimensional chromosome (the genotype).

Mutation. The mutation operator performs random modifications on an individual, i.e. changes the routing solution randomly. The purpose is to overcome local optima and to exploit new regions of the search space (Lienig and Thulasiraman 1994).

Reduction. The reduction strategy combines the current subpopulation with the newly created set of offspring, it then simply chooses the fittest individuals from the combined set to be the subpopulation in the next generation, thus keeping the subpopulation size constant.

15.4.3 Parallel structure

The island model used for this application has nine subpopulations ($N = 9$) connected by a torus topology. Thus, each subpopulation has exactly four *neighbors*. The total population ($M = 450$) is evenly partitioned ($\mu = 50$). (The listed numbers will be changed in the course of the experiments discussed in the following.) The parallel algorithm has been implemented on a network of SPARC workstations (SunOS and Solaris systems). The parallel computation environment is provided by the Mentat system, an object-oriented parallel processing system (Grimshaw 1993, Mentat 1996). The program, written in C++ and Fortran, comprises approximately 10 000 lines of source code. The cost factors in (15.1) are set to $w_1 = 1.0$, $w_2 = 2.0$, and $w_3 = 0.01$. The experimental results were achieved with the machines running their normal daily loads in addition to this application.

15.4.4 Comparison to other routing algorithms

Any experiment involving a 'real' application of an evolutionary algorithm should begin with a comparison to solution techniques that have already been acknowledged as effective by that application's community. Here we will simply state the results of the comparison because the detailed numbers will only be meaningful to the VLSI routing community and have appeared elsewhere (Lienig 1996). First, 11 benchmark problem instances were selected for channel and switchbox routing problems, for example, *Burstein's difficult channel* and *Joo6_16* for channels and *Joo6_17* and *Burstein's difficult switchbox* for switchboxes. These benchmarks were selected because published results were available for various routing algorithms, namely, Yoshimura and Kuh (1982), WEAVER (Joobbani 1986), BEAVER (Cohoon and Heck 1988), PACKER (Gerez and Herrmann 1989), SILK (Lin *et al* 1989), Monreale (Geraci *et al* 1991), SAR (Acan and Ünver 1989), and PARALLEX (Cho *et al* 1994). Note that most of these systems implement deterministic routers.

The island model was run 50 times per benchmark (with varying parameters) and the best-seen solution for each benchmark recorded. A comparison of those best-seen solutions to the previously published best-known solutions indicates that the island model solutions are qualitatively equal to or better than the best-known solutions from channel and switchbox routers for these benchmarks.

Of course, due to the stochastic nature of a GA, the best-seen results of the island model were not achieved in every run. (All executions were based on arbitrary initializations of the random number generator.) Above we refer to the *best-seen* solutions over all the runs for each benchmark; however, we would like to note that, in fact, in at least 50% of the individual island model runs solutions equal to these best-seen results were obtained. We judge this to be very consistent behavior for a GA.

15.5 The influence of island model parameters on evolution

Several experiments have been performed to illustrate the specific effects of important island model parameters in order to guide further applications of coarse-grained parallel GAs. The specific parameters varied in the experiments are the magnitude of migration, the frequency of migration, the epoch termination criterion, the migrant selection strategy, and the number of subpopulations and their sizes.

Five benchmark problem instances, namely, *Burstein's difficult channel*, *Joo6_13* and *Joo6_16* for channels, and *Joo6_17* and *pedagogical switchbox* for switchboxes, were chosen for these experiments. Comparisons were made between various parameter settings for the island model. In addition, runs were made with a sequential genetic algorithm (SGA) and a strictly isolated island model, i.e. no migration. The SGA executed the same algorithm as the subpopulation processes, but with a population size equal to the sum over all subpopulation sizes, i.e. $N * \mu$.

In the experiments, the SGA was set to perform the same number of recombinations per generation as the island model does over all subpopulations, namely, (number of subpopulations) × (offspring per subpopulation). The SGA and island model configurations were run for the same total number of generations. Thus, we ensure a fair method (with regard to the total number of solutions generated) to compare our parallel approach with an SGA.

The fundamental baseline was a derived measure based on the best-known objective measure for each problem instance. Remember that for the objective function, smaller values indicate higher-quality solutions. The derived measure is referred to as Δ and is calculated as indicated in the following. Let R_{bk} be the objective measure of the best-known solution and let R_{SGA} be the best-seen result on a particular run of SGA. Then,

$$\delta_{SGA} = \frac{R_{SGA} - R_{bk}}{R_{bk}} \tag{15.2}$$

is a relative (to the best-known) difference for a single run of SGA. The δ_{SGA} were averaged over five runs for each benchmark and over the five benchmarks to yield Δ_{SGA}.

In figures 15.4 and 15.6–15.8, this Δ_{SGA} is shown as a 100% bar in the leftmost position in the plot. Similar Δ values were obtained for the various island model configurations and are shown as percentages of Δ_{SGA}. Thus, if a particular island model configuration is shown with a 70% bar, then the average relative difference for that configuration is 30% better than SGA's average relative difference from the best-known result.

This derived measure was used in order to combine comparisons across problem instances with disparate objective function value ranges. In addition, the measure establishes a baseline through both best-known and SGA results. We remind the reader that for each benchmark problem this island model system evolved a solution equal to or better than any previously published system.

15.5.1 Number of migrants and epoch lengths

We investigated the influence of different epoch lengths (number of generations between migration) for different numbers of migrants (number of individuals sent to each of the four neighbors). The migrants were chosen randomly, with each migrant allowed to be sent only once. Figure 15.4 shows that the sequential approach was outperformed by all parallel variations (when averaged over all considered benchmarks). Note that the set of parallel configurations included the version with no migration, i.e. the strictly isolated island model (shown in

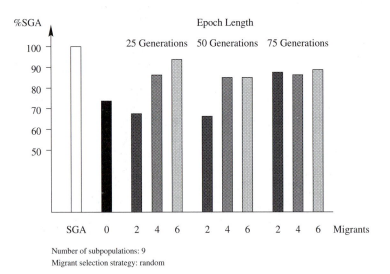

Figure 15.4. A comparison of results on the benchmark suite with different numbers of migrants and epoch lengths. Each bar is a Δ value for a different configuration. A Δ value is an average relative difference from the best-known solution as normalized to the Δ_{SGA} value, so the SGA bar is always 100%. Thus, the lower the bar, the better the average result of the particular configuration.

figure 15.4 as '0 migrants'). Thus, the splitting of the total population size into independent subpopulations has already increased the probability that at least one of these subpopulations will evolve toward a better result (given at least a 'critical mass' at each subpopulation, as we discuss in subsection 15.5.4).

Figure 15.4 also shows that a *limited* migration between the subpopulation further enhances the advantage of a parallel genetic algorithm. Two migrants to each neighbor with an epoch length of 50 generations are seen to be the best parameters when averaged over all problem instances. On the one hand, more migrants or too short epoch lengths are counterproductive to the idea of disjointly and parallel evolving subpopulations. The resulting intermixing diminishes the genetic diversity between the subpopulations by 'pulling' them all into the same part of the search space, thereby approaching the behavior of a single-population genetic algorithm. On the other hand, insufficient migration (an epoch length of 75 generations) simulates the isolated parallel approach (zero migrants)—the genetic richness of the neighboring subpopulations does not have enough chance to spread out.

Figure 15.5 shows this behavior in the context of individual subpopulations; that is, it presents the convergence behavior of the best individuals in each of the parallel evolving subpopulations on a specific problem instance (channel Joo6_13). It clearly indicates the importance of migration to avoid premature

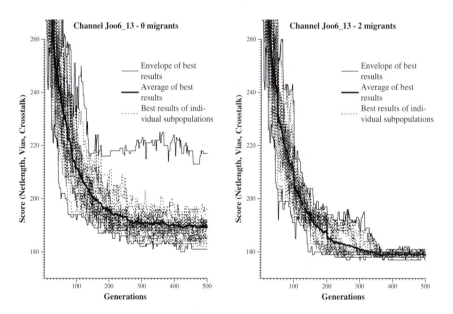

Figure 15.5. A comparison of the convergence of the best solutions in the individual, parallel evolving subpopulations. Plotted are five runs with nine subpopulations, i.e. 45 runs, in isolation (left) and with two migrants (right). (Note that the envelope for the plot on the left looks unusual due to an outlier subpopulation.)

stagnation by infusing new genetic material into a stagnating subpopulation. The 'stabilizing' effect of migration is also evident in the reduced variation among the best objective values gained in five independent runs, as shown in the right-hand plot of figure 15.5.

15.5.2   Variable epoch lengths

The theory of punctuated equilibria is based on two main ideas: (i) an isolated subpopulation in a constant environment will stabilize over time with little motivation for further development and (ii) continued evolution can be obtained by introducing new individuals from other, also stagnating subpopulations. However, all known computation models that are based on this theory use a fixed number of generations between migration. Thus, they do not exactly duplicate the model that migration occurs only *after* a stage of equilibrium has been reached within a subpopulation.

The algorithm was modified to investigate the importance of this characteristic. Rather than having a fixed number of generations between migrations, a stop criterion was introduced that took effect when stagnation in the convergence behavior within a subpopulation had been reached. After some experimentation with different models, we defined a suitable stop criterion to be 25 generations with no improvement in the best individual within a subpopulation.

To ensure a fair comparison, we kept the overall number of generations the same as in all other experiments. This led to varying numbers of epochs between the parallel evolving subpopulations (due to different epoch lengths) and resulted in longer overall completion time.

The results achieved with this variable epoch length are shown in figure 15.6. The results suggest that a slight improvement compared with a fixed epoch length can be achieved by this method. However, it is important to note that this comparison is made with a fixed epoch length that has been shown to be the most suitable after numerous experiments (see figure 15.4). Thus, the important attributes to notice are that the variable-epoch-length configuration frees the user from finding a suitable epoch length and that it gave more consistent results over the various migration settings.

15.5.3   Different migrant selection strategies

The influence of the quality of the migrants on the routing results was investigated using three migrant selection strategies: 'random' (migrants were chosen randomly with uniform distribution among the entire subpopulation), 'top 50%' (migrants were chosen randomly among the individuals with a fitness above the median fitness of the subpopulation), and 'best' (only the best individuals of the subpopulation migrated). The migrants were sent in a random order to the four neighbors.

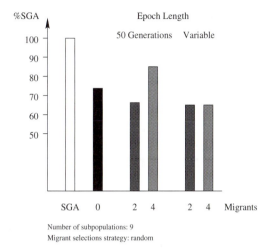

Figure 15.6. A comparison of results on the benchmark suite with fixed and variable epoch lengths. Variable-length epochs were terminated after 25 generations of no improvement of the best individual within the subpopulation. Each bar is a Δ value.

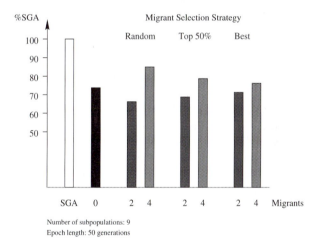

Figure 15.7. A comparison of results on the benchmark suite with different migrant selection strategies. Each bar is a Δ value.

As figure 15.7 indicates, we cannot find any improvement in the obtained results by using migrants with better quality. On the contrary, selecting better (or the best) individuals to migrate leads to a faster convergence—the final results are not as good as those achieved with a less elitist selection strategy. According to our observations, this is due to the dominance of the migrants having their presently superior genetic material reach all the subpopulations, thus leading the subpopulation searches into the same part of the search space concurrently.

15.5.4 Different numbers of subpopulations

To compare the influence of the number of subpopulations, the sizes of the subpopulations were kept constant and the number of subpopulations increased from $N = 9$ to $N = 16$ and $N = 25$ (still connected in a torus). Accordingly, the population size and the number of recombinations of the SGA were increased to maintain a fair comparison. The resulting plots for SGA against 16 and 25 subpopulations (with $\mu = 50$) are qualitatively similar to the SGA against nine subpopulations comparison of figure 15.4. For problems of this difficulty one would expect the SGA with slightly larger populations to do better than the small-population SGAs. That expectation was indeed born out in these experiments. The important observation is that the island model performance also increased; thus the relative performance advantage of the island model was maintained.

In an interesting variation on this experiment, the total population size, M, was held constant while increasing the number of subpopulations (and so reducing the subpopulation sizes). Holding M near 450 and increasing N to 16 yielded $\mu = 28$, while increasing N to 25 yielded $\mu = 18$.

The results presented in figure 15.8 show that subpopulation size is an important factor. The figure clearly indicates that (for this average measure)

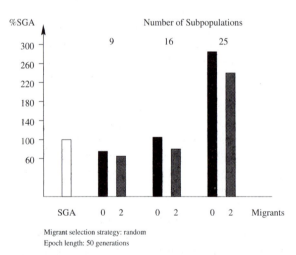

Figure 15.8. A comparison of results on the benchmark suite with different numbers of subpopulations. The size of the total population, M, is kept constant at 450 individuals. Since $M = N * \mu$, the increase in N, the number of subpopulations, requires a reduction in μ, the size of each subpopulation.

partitioning a population into subpopulations yielded (for $N = 9$) results better than SGA, then yielded progressively worse results as N was increased (and μ was decreased). We contend that this progression was due to the subpopulation size, μ, falling below 'critical mass' for the specific benchmark problem instances. Remember, the plotted values are aggregate measures. When we looked at the component values for each problem instance we found further evidence for our contention. The evidence was that for the simpler benchmarks the $N = 25$ island model still had extremely good performance, in fact, equaled the best-known solution repeatedly. For the other, more complex benchmarks, the $N = 25$ island model performed very poorly, thus dragging the average measure well below the SGA performance level. Thus, the advantage of more varied evolving subpopulations can be obtained by increasing N only if μ remains above 'critical mass'. This critical value for μ is dependent on the complexity of the problem instance being solved.

15.6   Final remarks and conclusions

In general it is difficult to compare sequential and parallel algorithms, particularly for stochastic processes such as GAs. As mentioned at the beginning of our discussion, the comparison is often according to overall time to completion, i.e. wall clock time. We contend that the island model constitutes a different evolutionary algorithm, not just a faster implementation of the SGA, and one that yields qualitatively better results. Here we have argued for this contention from the point of view of biological evolution and in the context of a difficult VLSI design application.

Several of the experimental design decisions we made for the application experiments merit reiteration here. First, the application is an important problem with an extensive literature of heuristic systems that 'solve' the problem. Our derived baseline measure incorporated the best-known objective values from this literature (see (15.2)). For this particular VLSI design problem, most of the heuristic systems are deterministic: thus we have aggregate values for the various GAs versus single values for the deterministic systems. Our measure does not directly account for this but we have provided an indication of the variation associated with the set of runs for the island model.

Second, our comparisons to an SGA are based on 'best-seen' objective values, not central processing unit (CPU) time or time to completion. In order to make these comparisons fair, we have endeavored to hold the computational resources constant and make them consistent across the SGA and the various configurations of the island model. This was done by fixing the total number of recombinations, i.e. the number of applications of the crossover operator, which relates directly to the total number of objective function evaluations.

Using the number of recombinations, as opposed to CPU time, for example, allows us to ignore properly implementation details for subsidiary processes such as sorting or insertion into a sorted list (Garey and Johnson 1979). Note that a CPU time measure can 'cut both ways' between the serial and parallel versions. On the one hand, if a subsidiary process has a small initial constant but poor performance as the data structure size increases, then the island model has an advantage simply through the partition of the population. On the other hand, if a subsidiary process is increasingly efficient for larger data structures but has a large initial constant, then the SGA has the advantage, again, simply through the partition of the population. These are examples of the subtle ways by which time-based comparisons confound primary search effort with irrelevant implementation details.

Third, our comparisons have ignored the cost of communication. This is generally appropriate for island models because the isolated computation time for each subpopulation is extremely large relative to the communication time for migration (any course-grained parallelization should have this attribute). For medium- and fine-grained parallel models communication is much more of a real concern. Ignoring communication time is also reflective of our interest in the evolutionary behavior of the models, not the raw speed of a particular implementation.

With all GAs, the evolutionary behavior depends heavily on the interplay between the problem complexity and population size. For SGA against island model comparisons this is particularly problematic because the island model has two population sizes: the total population size, M, and the subpopulation size, μ. Which is the proper one to consider relative to the SGA population size? For our experiments, we have used the total population size. We believe this makes the versions more conformable and is most consistent with number of recombinations as the computation measure.

Further, for comparing stochastic processes, such as GAs, the total number of trials is important, particularly when the evaluation measure is based on best-seen results. The attentive reader will have noticed that the strictly isolated island model (no migration) often does better than the SGA. This might seem curious, since a single run of the isolated island model is just a set of N separate SGAs and ones with smaller population sizes. As long as the subpopulation size, μ, is above what we are calling the 'critical mass' level, the isolated island model has a statistical advantage over the single SGA (under our evaluation measure).

The evaluation measure gives this advantage because the best seen is taken over each run. Thus, each run of the isolated island model has N 'samples' to determine its best seen, while each SGA run has only one 'sample'. (Shonkwiler (1993) calls these *IIP parallel* GAs and gives an analysis of 'hitting time' expectation.) We consider the 'samples' in this case to be evolutionary trajectories through the solution space. Now, note that in almost all cases allowing migration provides the island model with the means to derive even better results.

This application of the island model to detailed routing problems in VLSI circuit design has shown that a parallel GA based on the theory of *punctuated equilibria* outperforms an SGA. Furthermore, the results are qualitatively equal to or better than the best-known results from published channel and switchbox routers.

In investigating the parameters of the island model, the following conclusions have been reached:

- The island model consistently performs better than the SGA, given a consistent amount of computation.

- The size of a subpopulation, the total amount of immigration, i.e. the number of connected subpopulations multiplied by the number of migrants per neighbor, the epoch length and the complexity of the problem instance are interrelated quantities. The problem instance complexity determines a minimum population size for a viable evolutionary trajectory. The total amount of immigration must not be disruptive (our experiments indicate that more than 25% of the subpopulation size is disruptive), and the epoch length must be long enough to allow exploitation of the infused genetic material. Within these constraints the island model will perform better with more subpopulations, even while holding the total population size and the total number of recombinations, i.e. amount of computation, constant.

- Variable epoch lengths determined via equilibrium measures within subpopulations achieve overall results slightly better than those obtained with (near-) optimized fixed epoch lengths. Though an equilibrium measure must be chosen, allowing variable epoch lengths frees the user from having to select this parameter value.

- Quality constraints on the migrants do not improve the overall behavior of the algorithm: on the contrary, quality requirements on the selection of the migrants increases the occurrence of premature stagnation.

- Given a sufficient number of individuals per subpopulation, the larger the number of parallel evolving subpopulations, the better the routing results. The complexity of the problem and the minimal subpopulation size have a direct correlation that must be taken into account when dividing a population into subpopulations.

Finally, we would like to return to an issue that we mentioned at the very beginning of our discussions: namely, the island model formulation of the GA is not simply a hardware accelerator of the single-population GA. The island model does map naturally to distributed-memory message-passing multiprocessors, so it is amenable to the speedup in time to completion that such parallel architectures can provide. However, the formulation can improve the quality of solutions obtained even via sequential simulations of the island model. As supported by the *shifting balance* and *punctuated equilibria* theories, the emergent properties of the computation derive from the *concurrent* evolutionary trajectories of the subpopulations *interacting* through limited migration.

References

Acan A and Ünver Z 1992 Switchbox routing by simulated annealing: SAR *Proc. IEEE Int. Symp. on Circuits and Systems* vol 4, pp 1985–8

Adamidis P 1994 *Review of Parallel Genetic Algorithms* Technical Report, Department of Electrical and Computer Engineering, Aristotle University, Thessaloniki

Bailey J 1992 First we reshape our computers, then our computers reshape us: the broader intellectual impact of parallelism *Daedalus* pp 67–86

Barnes G H, Brown R M, Kato M, Kuck D J, Slotnick D L and Stokes R A 1968 The ILLIAC IV computer *IEEE Trans. Computer* **C-17** 746–57

Cho T W, Sam S, Pyo J and Heath R 1994 PARALLEX: a parallel approach to switchbox routing *IEEE Trans. Computer-Aided Design* **CAD-13** 684–93

Cohoon J P and Heck P L 1988 BEAVER: a computational-geometry-based tool for switchbox routing *IEEE Trans. Computer-Aided Design* **CAD-7** 684–97

Cohoon J P, Hegde S U, Martin W N and Richards D S 1987 Punctuated equilibria: a parallel genetic algorithm *Proc. 2nd Int. Conf on Genetic Algorithms (Pittsburgh, PA, 1987)* ed J J Grefenstette (Hillsdale, NJ: Erlbaum) pp 148–54

——1991a Distributed genetic algorithms for the floorplan design problem *IEEE Trans. Computer-Aided Design* **CAD-10** 483–92

Cohoon J P, Martin W N and Richards D S 1991b Genetic algorithms and punctuated equilibria in VLSI *Parallel Problem Solving from Nature (Lecture Notes in Computer Science 496)* ed H P Schwefel and R Männer (Berlin: Springer) pp 134–44

Eldredge N 1989 *Macro-evolutionary Dynamics: Species, Niches, and Adaptive Peaks* (New York: McGraw-Hill)

Eldredge N and Gould S J 1972 Punctuated equilibria: an alternative to phyletic gradualism *Models of Paleobiology* ed T J M Schopf (San Francisco, CA: Freeman, Cooper) pp 82–115

Forrest S (ed) 1991 *Emergent Computation* (Cambridge, MA: MIT Press)

Fung L W 1976 *MPPC: a Massively Parallel Processing Computer* Goddard Space Flight Center Section Report

Garey M R and Johnson D S 1979 *Computers and Intractability: a Guide to the Theory of NP-completeness* (San Francisco, CA: Freeman)

Geraci M, Orlando P, Sorbello F and Vasallo G 1991 A genetic algorithm for the routing of VLSI circuits *Proc. Euro ASIC '91* pp 218–23

Gerez S H and Herrmann O E 1989 Switchbox routing by stepwise reshaping *IEEE Trans. Computer-Aided Design* **CAD-8** 1350–61

Goldberg D E 1989 *Genetic Algorithms in Search, Optimization, and Machine Learning* (Reading, MA: Addison-Wesley)

Gordon V S, Whitley D and Böhn A 1992 Dataflow parallelism in genetic algorithms *Parallel Problem Solving from Nature, 2 (Brussels 1992)* ed R Männer and B Manderick (Amsterdam: Elsevier Science) pp 533–42

Gorges-Schleuter M 1990 Explicit parallelism of genetic algorithms through population structures *Parallel Problem Solving from Nature (Lecture Notes in Computer Science 496)* ed H-P Schwefel and R Männer (Berlin: Springer) pp 150–9

Gorges-Schleuter M 1992 Comparison of local mating strategies *Parallel Problem Solving from Nature, 2 (Brussels 1992)* ed R Männer and B Manderick (Amsterdam: Elsevier Science) pp 553–62

Grimshaw A S 1993 Easy-to-use object-oriented parallel programming with Mentat *IEEE Computer* **26** 39–51

Grosso P 1985 *Computer Simulation of Genetic Adaptation: Parallel Subcomponent Interaction in a Multilocus Model* PhD Thesis, Computer and Communication Sciences Department, University of Michigan

Holland J H 1975 *Adaptation in Natural and Artificial Systems* (Ann Arbor, MI: University of Michigan Press)

Huxley J 1942 *Evolution: the Modern Synthesis* (New York: Harper)

Joobbani R 1986 *An Artificial Intelligence Approach to VLSI Routing* (Boston, MA: Kluwer)

Lienig J 1996 A parallel genetic algorithm for two detailed routing problems *IEEE Int. Symp. on Circuits and Systems (Atlanta, GA, 1996)* pp 508–11

Lienig J and Thulasiraman K 1994 A genetic algorithm for channel routing in VLSI circuits *Evolutionary Comput.* **1** 293–311

Lin Y-L, Hsu Y-C and Tsai F-S 1989 SILK: a simulated evolution router *IEEE Trans. Computer-Aided Design* **CAD-8** 1108–14

Lohmann R 1990 Application of evolution strategies in parallel populations *Parallel Problem Solving from Nature (Lecture Notes in Computer Science 496)* ed H P Schwefel and R Männer (Berlin: Springer) pp 198–208

Mentat 1996 homepage http://www.cs.virginia.edu/~mentat/

Mühlenbein H 1989 Parallel genetic algorithms, populations genetics and combinatorial optimization *Proc. 3rd Int. Conf on Genetic Algorithms (Fairfax, VA, 1989)* ed J D Schaffer (San Mateo, CA: Morgan Kaufmann) pp 416–21

Mühlenbein H, Gorges-Schleuter M and Krämer O 1987 New solutions to the mapping problem of parallel systems—the evolution approach *Parallel Comput.* **6** 269–79

Pettey C B, Leuze M R and Grefenstette J J 1987 A parallel genetic algorithm *Proc. 2nd Int. Conf. on Genetic Algorithms (Pittsburgh, PA, 1987)* ed J J Grefenstette (Hillsdale, NJ: Erlbaum) pp 155–61

Radcliffe N J 1991 Forma analysis and random respectful recombination *Proc. 4th Int. Conf on Genetic Algorithms (San Diego, CA, 1991)* ed R K Belew and L B Booker (San Mateo, CA: Morgan Kaufmann) pp 222–9

Rudolph G 1990 Global optimization by means of distributed evolution strategies *Parallel Problem Solving from Nature (Lecture Notes in Computer Science 496)* ed H P Schwefel and R Männer (Berlin: Springer) pp 209–13

Seitz C L 1985 The cosmic cube *Commun. ACM* **28** 22–33

Shonkwiler R 1993 Parallel genetic algorithms *5th Int. Conf. on Genetic Algorithms (Urbana-Champaign, IL, 1993)* ed S Forrest (San Mateo, CA: Morgan Kaufmann) pp 199–205

Slotnick D, Borck W and McReynolds R 1962 The SOLOMON computer *Proc. Fall Joint Computer Conf.* (AFIPS) pp 97–107

Tamaki H and Nishikawa Y 1992 A parallel genetic algorithm based on a neighborhood model and its application to the jobshop scheduling *Parallel Problem Solving from Nature, 2 (Brussels 1992)* ed R Männer and B Manderick (Amsterdam: Elsevier Science) pp 573–82

Tanese R 1987 Parallel genetic algorithms for a hypercube *Proc. 2nd Int. Conf on Genetic Algorithms (Pittsburgh, PA, 1987)* ed J J Grefenstette (Hillsdale, NJ: Erlbaum) pp 177–83

Wright S 1932 The roles of mutation, inbreeding, crossbreeding and selection in evolution *Proc. 6th Int. Congr. Genetics* vol 1, pp 356–66

——1964 Stochastic processes in evolution *Stochastic Models in Medicine and Biology* ed J Gurland (Madison, WI: University of Wisconsin Press) pp 199–241

——1982 Character change, speciation, and the higher taxa *Evolution* **36** 427–43

Wulf W A and Bell C G 1972 C.mmp—a multi-mini-processor *Proc. Fall Joint Conf* (AFIPS) pp 765–77

Yoshimura T and Kuh E S 1982 Efficient algorithms for channel routing *IEEE Trans. Computer-Aided Design* **CAD-1** pp 25–35

16

Diffusion (cellular) models

Chrisila C Pettey

16.1 A formal description of the diffusion model

Since a generation is the parallel evolution of μ individuals, the diffusion model is implemented by placing one individual per processor. Thus, μ is determined by and is equal to the number of processing nodes in the parallel environment. Because of this limitation, the population size remains constant at all times. Furthermore, since the idea of a generation involves mating, the diffusion model generally contains some form of recombination. Given these two limitations of fixed population size and recombination, the diffusion model is almost always a genetic algorithm (GA) model.

In the diffusion model, each process is responsible for evolving its individual. In keeping with the deme theory, while selection and recombination could be performed globally, they generally are performed within a local neighborhood. Thus, the pseudocode for a single process in this model might appear as follows:

```
Process (i):
    1. t ← 0;
    2. initialize individual i;
    3. evaluate individual i;
    4. while (t ≤ t_max) do
    5.      individual i ← select(neighborhood(i))
    6.      choose parent 1 from neighborhood
    7.      choose parent 2 from neighborhood
    8.      individual i ← recombine(parent1, parent2)
    9.      individual i ← mutate(individual i)
    10.     evaluate individual i;
    11.     t ← t + 1;
         od
```

It has been proved by Whitley (1993) that any evolutionary algorithm (EA) of this form is equivalent to a cellular automaton. Thus, the diffusion model could also be called the cellular model. As in all EA models, in order to implement this model, it will be necessary to determine what an individual will look like, how an individual will be initialized and evaluated, what will be the mutation rate, and what will be the stopping criteria. On the other hand, the key implementation issues that are unique to the diffusion model—how should selection be performed, how are the parents chosen, what is the size and shape of the neighborhood, and how is recombination performed—along with several techniques for implementing each of these issues are presented in the remainder of this chapter. It should be noted, that since evolution is viewed on an individual basis, throughout the rest of this section the diffusion model will be viewed from the perspective of a single process which is evolving a single individual.

16.2 Diffusion model implementation techniques

When considering how to implement the diffusion model, it is tempting to perform a global selection and recombination because then the model would theoretically be equivalent to a sequential EA. This solution is made even more inviting by the fact that all massively parallel machines have a front-end processor or a host processor for handling global operations. The problem with this solution is that in a parallel machine environment global operations are much more time consuming than local operations. The problem is magnified in a distributed environment. To overcome the inefficiency of the global solution and to maintain the idea of the individual being the parallel part of the process, almost all implementations (for a counterexample see Farrell *et al* (1994)) perform selection and recombination in the local neighborhood (Collins and Jefferson 1991). Furthermore, since a process is only responsible for its individual, selection can be combined with the process of choosing the parents for recombination. In most implementations that perform selection and recombination in separate steps, the selection is a local implementation of a global technique such as proportional, ranking, or binary tournament selection (see for example Collins and Jefferson 1991, De Jong and Sarma 1995). The local versions of the global techniques are implemented just like the global techniques with the exception that the individual's population is just the neighborhood—not the global population. Selection will not be discussed here. In this section, techniques for choosing parents, neighborhood size and shape attributes, and recombination techniques are presented. However, since many techniques and choices may be based on the parallel environment, it is necessary to begin with a short description of typical parallel environments for the diffusion model.

16.2.1   Parallel environments for diffusion model implementation

Michael Flynn (1966) coined the terms MIMD (multiple instruction–multiple data) and SIMD (single instruction–multiple data), which are used to characterize parallel processing paradigms. Both terms are typically used to characterize hardware paradigms, but, perhaps just as typically, the terms are also applied to the algorithms that are written to exploit the underlying hardware. A third term, SPMD (single program–multiple data), is a term that is applied to the situation where the algorithm is a SIMD algorithm, but the underlying hardware is MIMD.

In a MIMD environment, processors work independently and asynchronously. Coordination and communication among processes in the form of locks, barriers, shared variables, or message passing are handled by the programmer. Some examples of MIMD machines are hypercubes, Thinking Machines CM-5, and the Intel Paragon. Another MIMD environment that is becoming more common is clusters of workstations (or farms) running a parallel software system such as PVM, Linda, or Express. A MIMD program consists of two or more (usually different) processes. In this sense, it is the functions which are executed in parallel. Thus, this form of parallelism is usually called functional (or course-grained) parallelism. With the advent of the hypercube in the mid-1980s, the MIMD computing environment became accessible to more people. As a result, the island (migration) model (Chapter 15), which is well suited to a MIMD environment, was more popular than the diffusion model.

In the last decade, however, SIMD machines have become more readily available and more user friendly. Some typical SIMD machines are Thinking Machines CM-1 and CM-2, Active Memory Technology Ltd DAP, and MasPar MP1 and MP2. All SIMD machines have a front-end processor for handling the code and the global operations. Behind the front-end processor is the processor array consisting of sometimes thousands of processors. In most cases, the interconnection network of the processor array is a mesh (or grid). In a SIMD environment, processors work synchronously, that is, in lock step. Data are partitioned across the processors, and each processor performs the same instruction on its data. Process coordination is handled in the hardware, and process communication is usually handled with built-in primitives. In a SIMD program, multiple processors execute the same process at the same time. Since each processor has different data, it is the data transformations that are being performed in parallel. Thus, this form of parallelism is usually called data (or fine-grained) parallelism.

The SIMD environment is perfect for the diffusion model. However, if the only available environment is a MIMD environment, it is still possible to implement data parallelism, and, thus, the diffusion model. In the diffusion model, it is not necessary for the evolution of the individual to take place in lock step. Each individual can evolve independently of the others much as in nature. Therefore, it is possible to use the SPMD programming paradigm. In a SPMD environment, each processor runs the same program on different data. Because the underlying hardware is MIMD, the processes will run asynchronously. The programmer is responsible for handling process coordination and communication using locks, barriers, shared variables, or message passing. In this situation, the easiest solution would be to allow the individuals to evolve asynchronously. However, using barriers, it is possible to force the processes to wait on all other processes before proceeding with the next generation.

With a basic understanding of data parallelism, it is now possible to continue with the discussion of the implementation issues in the diffusion model. For additional information on functional parallelism, data parallelism, MIMD machines or environments, or SIMD machines see the books by Almasi and Gottlieb (1994) or Morse (1994).

16.2.2 Techniques for selecting parents

No matter what technique is used for selecting the parents, it will be necessary for each process to communicate with some (maybe all) of its neighbors. If possible, communication in a parallel environment should be kept to a minimum in order to improve the performance of the algorithm. Therefore, the neighborhood sizes are usually kept small in order to alleviate the communication problem.

Quite frequently the technique for selecting the parents for recombination is a local version of a standard global selection technique. When converting a global technique to a local technique it is simply a matter of implementing the global technique as if it were in a much smaller population. This will mean collecting performance measures from all individuals in the neighborhood, or as in the case of *tournament selection* collecting the performance measures from a random set of individuals in the neighborhood.

There are many examples of local implementations of global techniques. For example, Gorges-Schleuter (1989), Manderick and Spiessens (1989), and Davidor (1991) all use proportional selection for at least one parent. The difference between the three techniques is in the selection of the second parent. Gorges-Schleuter chooses the process' individual as the second parent. Manderick and Spiessens choose the second parent randomly. Davidor chooses both parents based on the probability distribution of the neighborhood fitnesses.

Other examples of selection techniques in diffusion model implementations may involve as much communication as the previously mentioned techniques. For example, Farrell et al (1994) have devised one such technique. The first parent chosen is the individual. The second parent is the most successful individual in the neighborhood.

One technique which may or may not involve as much communication is the technique devised by Collins and Jefferson (1991). In their diffusion model, each parent is selected by performing a random walk. The fittest individual found in the random walk becomes the parent. Of course the length of the random walk determines the amount of communication.

Probably the most unique technique is to choose all of the neighbors as parents. Mühlenbein (1989) chose the four neighbors, the individual, and the global best individual was chosen twice (i.e. there were seven parents!). Of course this type of selection affects the recombination. Recombination techniques are discussed in the following section.

16.2.3 Recombination techniques

In most cases the recombination technique is a 'typical' technique. In other words, the technique is the same as a sequential EA. One interesting deviation from the typical techniques is p-sexual voting (Mühlenbein 1989). As was mentioned previously, in this diffusion model, each child has more than two parents. All parents vote on which allele should be chosen for a particular gene. If one allele receives more than some threshold number of votes, then that allele wins. Otherwise an allele is chosen randomly for the gene.

Regardless of the recombination technique, the number of children created by recombination can be one or two as in sequential EAs. Also as in sequential EAs, the question arises as to what should be done with the child(ren). If only one child is created, it usually replaces the individual, although Gorges-Schleuter (1989) only allows a replacement to occur if the fitness of the child is not the worst fitness in the neighborhood. If two children are created, usually one is chosen (see e.g. Manderick and Spiessens 1989) and it replaces the individual. However, Davidor (1991) creates two children and places both in the neighborhood. If the child is different from the individual which it is supposed to replace, then one of the two is selected based on the probability distribution created by their two fitnesses.

Often the selection technique and the replacement of children is influenced by the size and shape of the neighborhood. Some typical deme attributes are presented in the following section.

16.2.4 Deme attributes: size and shape

In most diffusion model implementations, the size and shape of the deme is influenced by the underlying architecture. For instance, the most common SIMD interconnection topology is a mesh or grid. Given this underlying architecture, the typical selection for a neighborhood is the neighboring processors on the grid. In many of these machines it is also possible to quickly communicate with the NE, NW, SE, and SW neighbors as well as the neighbors immediately above, below, to the left, and to the right of the processor. This creates a neighborhood of nine individuals. For example, in figure 16.1 the circles represent processors, and the lines represent connections between processors. The neighborhood of the individual residing on the processor represented by the open circle would be all nine circles in the figure (Manderick and Spiessens 1989, Davidor 1991).

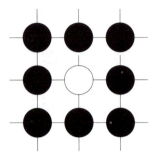

Figure 16.1. A mesh (or grid) neighborhood.

The underlying architecture used by Mühlenbein (1989) and Gorges-Schleuter (1989) was a double ring. Each processor in each ring was connected to exactly one processor in the other ring. This produced a ladder-like topology. In this situation, an individual's neighborhood can be determined by how close a processor is to other processors on the ladder. Closeness is usually defined in terms of how many communication links away a processor is from a given processor. For instance, if the neighborhood is defined as being all processors no more than one link away, then the neighborhood is T shaped with the individual, the individual to the left on the ring, the individual to the right on the ring, and the neighboring individual on the other ring (see figure 16.2). Figure 16.3 shows the neighborhood if the distance between neighbors is two.

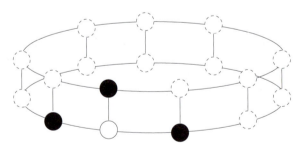

Figure 16.2. A ladder neighborhood with a distance of one.

Occasionally the selection technique and not the architecture affects the size and shape of the neighborhood. Collins and Jefferson (1991) implemented their diffusion model on a grid, but because the selection technique was a random walk, then the selection technique affected the deme shape and size. The size of the deme was the length of the random walk. The shape of the deme was random.

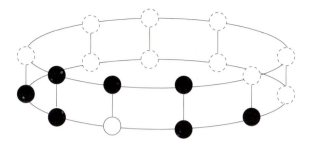

Figure 16.3. A ladder neighborhood with a distance of two.

In this section several implementation techniques have been presented. It should be noted that most techniques cause a theoretical deviation from the underlying sequential EA theory. In the next section a few remarks are made about theoretical research in the area of diffusion models.

16.3 Theoretical research in diffusion models

Very little has been done theoretically in the area of diffusion models. This is probably due in part to the difficulty of deriving generic proofs in an area where there are so many different implementation possibilities. Another possibility for the lack of theory may be the belief that the diffusion model more correctly models natural populations than the island model or sequential EAs. Regardless of the reason for the lack of theory, more work needs to be done in this area. Below are mentioned three of the theoretical results that have been published.

Davidor (1991) derived a schema theorem for eight neighbors on a grid. The theorem was based on using proportional selection for both parents, creating two children, and proportionally placing both children in the neighborhood. Using this theorem, he found that the diffusion model had a rapid but local convergence creating 'islands of near optimal strings'. This rapid, local convergence is no surprise considering that the selection is effectively in very small populations.

Spiessens and Manderick (1991) performed a comparison of the time complexity of their diffusion model and a sequential GA. They ignored the evaluation step since this is problem dependent. They were able to show that the complexity of the diffusion model increases linearly with respect to the length of the genotype. The complexity of a sequential GA increases polynomially with respect to the size of the population multiplied by the length of the genotype. Since, theoretically, an increase in the length of an individual should be accompanied by an increase in the population size, an increase in the length of an individual will affect the run time of a sequential GA, but it will not affect the run time of the diffusion model. Also in this article, they derive the expected number of individuals due to proportional, scaling, local ranking, and local tournament selection. The growth rates are then compared showing that proportional selection has the lowest growth rate.

The final result presented here was actually a result of experiments done by De Jong and Sarma (1995). In their work they discovered a result that needs to be kept in mind by all who would implement the diffusion model. Their experiments compared proportional, ranking, and binary tournament selection. While performing their experiments they found that binary tournament selection appeared to perform worse than linear ranking. This was surprising given that the two techniques have equivalent selection pressures. 'These results emphasize the importance of an analysis of the variance of selection schemes. Without it one can fall into the trap of assuming that selection algorithms that have equivalent expected selection pressure produce similar search behavior' (De Jong and Sarma 1995).

16.4 Conclusion

The diffusion model is perhaps the most 'natural' EA in that it seems to simulate the evolution of natural populations from the point of view of the individual. While there are a few implementation issues that are unique to the diffusion model, it is, none the less, a fairly simple algorithm to implement. This, coupled with the increasing availability of SIMD and SPMD environments, makes the diffusion model an excellent choice for improving the search of an EA.

16.5 Additional sources of information

It was impossible to list all the authors or all the implementations of the diffusion model. Therefore, to avoid accidentally leaving someone out, only a few examples were chosen for the preceding sections and the bibliography. There are also many good Internet resources, but most of them can be reached from ENCORE (the evolutionary computation repository) on the World Wide Web.

References

Almasi G S and Gottlieb A 1994 *Highly Parallel Computing* (Redwood City, CA: Benjamin–Cummings)

Collins R J and Jefferson D R 1991 Selection in massively parallel genetic algorithms *Proc. 4th Int. Conf. on Genetic Algorithms (University of California, San Diego, CA, July 1991)* ed R K Belew and L B Booker (San Mateo, CA: Morgan Kaufmann) pp 249–56

Davidor Y 1991 A naturally occurring niche and species phenomenon: the model and first results *Proc. 4th Int. Conf. on Genetic Algorithms (University of California, San Diego, CA, July 1991)* ed R K Belew and L B Booker (San Mateo, CA: Morgan Kaufmann) pp 257–63

De Jong K and Sarma J 1995 On decentralizing selection algorithms *Proc. 6th Int. Conf. on Genetic Algorithms (Pittsburgh, PA, July 1995)* ed L J Eshelman (San Mateo, CA: Morgan Kaufmann) pp 17–23

Farrell C A, Kieronska D H and Schulze M 1994 Genetic algorithms for network division problem *Proc. 1st IEEE Conf. on Evolutionary Computation (Orlando, FL, June 1994)* (Piscataway, NJ: IEEE) pp 422–7

Flynn M J 1966 Very high speed computing systems *Proc. IEEE* **54** 1901–9

Gorges-Schleuter M 1989 ASPARAGOS: an asynchronous parallel genetic optimization strategy *Proc. 3rd Int. Conf. on Genetic Algorithms (Fairfax, VA, June 1989)* ed J D Schaffer (San Mateo, CA: Morgan Kaufmann) pp 422–7

Hartl D L 1980 *Principles of Population Genetics* (Sunderland, MA: Sinauer)

Manderick B and Spiessens P 1989 Fine-grained parallel genetic algorithms *Proc. 3rd Int. Conf. on Genetic Algorithms (Fairfax, VA, June 1989)* ed J D Schaffer (San Mateo, CA: Morgan Kaufmann) pp 428–33

Morse H S 1994 *Practical Parallel Computing* (Cambridge, MA: AP Professional)

Mühlenbein H 1989 Parallel genetic algorithms, population genetics and combinatorial optimization *Proc. 3rd Int. Conf. on Genetic Algorithms (Fairfax, VA, June 1989)* ed J D Schaffer (San Mateo, CA: Morgan Kaufmann) pp 416–21

Spiessens and Manderick 1991 A massively parallel genetic algorithm: implementation and first analysis *Proc. 4th Int. Conf. on Genetic Algorithms (University of California, San Diego, CA, July 1991)* ed R K Belew and L B Booker (San Mateo, CA: Morgan Kaufmann) pp 279–86

Whitley D 1993 Cellular genetic algorithms *Proc. 5th Int. Conf. on Genetic Algorithms (Urbana-Champaign, IL, July 1993)* ed S Forrest (San Mateo, CA: Morgan Kaufmann)

17

Population sizing

Robert E Smith

17.1 Introduction

How large should a population be for a given problem? This question has been considered empirically in several studies (De Jong 1975, Grefenstette 1986, Schaffer *et al* 1989), and there are a variety of heuristic recommendations on sizing populations for a variety of EC algorithms. This section considers analytically motivated suggestions on population sizing. It is primarily focused on genetic algorithms (GAs), since most of the analytical work on sizing populations is in this EC subfield. However, many of the concepts discussed can be transferred to other EC algorithms.

The issue of population sizing can also be considered theoretically in the light of GA schema processing. By misinterpreting the *implicit parallelism* ($O(\mu^3)$) argument, one might conclude that the larger μ (the population size) is, the greater the computational leverage, and, therefore, the better the GA will perform. This is clearly not the case, since there are only 3^ℓ schemata in a binary-encoded GA with strings of length ℓ. One clearly cannot process $O(\mu^3)$ schemata if μ^3 is much greater than 3^ℓ.

There are several ways to view the sizing of a GA population. One is to size the population such that computational leverage (i.e. schema processing ability) is maximized. Goldberg (1989) shows how to set population size for an optimal balance of computational effort to computational leverage. This development is outlined below. However, it is important to note that there are other GA performance criteria that can (and should) be considered when sizing a population. One is the accuracy of schema average fitness values indicated by a finite sample of the schemata in a population. This issue will be considered later in this chapter. However, put in a broader context, population sizing cannot be considered in complete isolation from other GA parameters. Ultimately, the GA must balance computational leverage, accurate sampling of schemata, population diversity, mixing through recombination, and selective pressure for good performance.

134

17.2   Sizing for optimal schema processing

To consider how the computational leverage of implicit parallelism can be maximized, one must thoroughly consider the number of schemata processed by the GA. One can derive an exact expected number of schemata in a population of size n given binary strings of length ℓ. Note that this argument can be extended to higher-cardinality alphabets as well.

First consider the probability that a single string matches a particular schema H:

$$P_H = \left(\tfrac{1}{2}\right)^{O(H)}.$$

Given this, the probability of zero matches of this schema in a population of size μ is

$$\left[1 - \left(\tfrac{1}{2}\right)^{O(H)}\right]^{\mu}.$$

Therefore, the probability of one or more matches in a population of size n is

$$1 - \left[1 - \left(\tfrac{1}{2}\right)^{O(H)}\right]^{\mu}.$$

There are

$$\binom{\ell}{O(H)} 2^{O(H)}$$

schemata of order $O(H)$ in a strings of length ℓ. Therefore, if one counts over all possible schemata, and considers the probability of one or more of those schemata in a population of size μ, the total, expected number of schemata in the population is

$$S(\mu, l) = \sum_{i=0}^{\ell} \binom{\ell}{i} 2^i \left\{1 - \left[1 - \left(\frac{1}{2}\right)^i\right]^{\mu}\right\}.$$

Consider schemata of defining length ℓ_s or less, such that these schemata are highly likely to survive crossover and mutation. These schemata can be thought of as building blocks. Given the previous count of schemata, one can slightly underestimate the number of building blocks as

$$n_s(\mu, \ell, \ell_s) = (\ell - \ell_s + 1)\, S(\mu, \ell_s - 1).$$

Note that the number of building blocks monotonically expands from $(\ell - \ell_s + 1)\, 2^{\ell_s}$ (for a population of one) to $(\ell - \ell_s + 1)\, 3^{\ell_s}$ (for an infinite population).

Given this count, a measure of the GA's computational leverage is dS/dt, the average real-time rate of schemata processing. Assume that the population ultimately converges to contain only one unique string, and, thus, 2^{ℓ} schemata. Therefore, for the overall GA run, one can estimate

$$\frac{dS}{dt} \approx \frac{(S_0 - 2^{\ell})}{Dt}$$

where S_0 is the expected number of unique schemata in the initial, random population (given by the S count equation above), and Dt is the time to convergence.

Assume

$$Dt = n_c t_c$$

where n_c is the generations to convergence, and t_c is the real time per generation. Goldberg (1989) estimates the convergence time under fitness proportionate selection. If one considers convergence of all but λ percent of the population to one string, where λ is the initial percentage of the population occupied by the string, the time to convergence is constant with respect to population size. If one considers convergence to all but one of the population members to the same string, the convergence time is $n_c = O(\ln \mu)$.

The time t_c varies with the degree of parallelization, since parallel computers can evaluate several fitness values simultaneously. Therefore,

$$t_c = \mu^{(1-\beta)}.$$

The value $\beta = 1$ represents a perfectly parallel computer, where all fitness values are evaluated at once. The value $\beta = 0$ represents a perfectly serial computer. Note that Dt increases monotonically with n.

Since DS/Dt is given as an analytical function, one can find its maxima using standard numerical search techniques. Goldberg (1989) compiles maxima for several values of ℓ_s and β in plots and tables.

Surprisingly, this development for serial computers and the $O(\ln \mu)$ convergence time assumption indicate that one should use the smallest population possible. A population of size three seems the smallest that is technically feasible, since two population members are required to be selected over the third, and then recombined to form a new population. If one starts with such small populations, convergence will be rapid, and then the GA can be restarted. This inspired the *micro-GA* (Krishnakumar 1989), which uses very small populations, and repeated, partially random restarts. Although results of the micro-GA are promising, it is important to note that the optimal population size for schema processing rate *may not* be the optimal size for ultimate GA effectiveness. Sampling error may overwhelm the GA's ability to select correctly in small populations.

17.3   Sizing for accurate schema sampling

Another study by Goldberg *et al* (1992) examines population sizing in terms of ultimate GA performance by considering sampling error. Basic GA theory suggests that GAs search by implicitly evaluating the mean fitness of various schemata based on a series of population samples, and then recombining highly fit schemata. Since the schema average fitness values are based on samples, they typically have a nonzero variance. Consider the competing schemata

$$H_1 = * \ * \ * \ * \ 1 \ 1 \ 0 \ * \ * \ 0$$
$$H_2 = * \ * \ * \ * \ 0 \ 1 \ 0 \ * \ * \ 0.$$

Assuming a deterministic fitness function, variance of average fitness values of these schemata exists due to the various combinations of bits that can be placed in the 'don't care' (*) positions. This variance has been called *collateral noise* (Goldberg and Rudnick 1991). Let $f(H_1)$ and $f(H_2)$ represent the average fitness values for schemata H_1 and H_2, respectively, taken over all possible strings in each schema. Also let σ_1^2 and σ_2^2 represent the variances taken over all corresponding schema members.

The GA does not make its selection decisions based on $f(H_1)$ and $f(H_2)$. Instead, it makes these decisions based on a sample of a given size for each schema. Let us call these observed fitness values $f_o(H_1)$ and $f_o(H_2)$. Observed fitness values are a function of $n(H_1)$ and $n(H_2)$, the numbers of copies of schemata H_1 and H_2 in the population, respectively. Given moderate sample sizes, the central-limit theorem† tells us that the f_o-values will be distributed normally, with mean $f(H)$ and variance $\sigma^2/n(H)$.

Due to the sampling process and the related variance, it is possible for the GA to err in its selection decisions on schema H_1 versus H_2. In other words, if one assumes $f(H_1) > f(H_2)$, there is a probability that $f_o(H_1) < f_o(H_2)$. If such mean fitness values are observed the GA may incorrectly select H_2 over H_1. Given the $f(H)$ and σ^2 values, one can calculate the probability of $f_o(H_1) < f_o(H_2)$ based on the convolution of the two normals. This convolution is itself normal with mean $f(H_1) - f(H_2)$ and variance $(\sigma_1^2/n(H_1)) + (\sigma_2^2/n(H_2))$. Thus, the probability that $f_o(H_1) < f_o(H_2)$ is α, where

$$z^2(\alpha) = \frac{(f(H_1) - f(H_2))^2}{(\sigma_1^2/n(H_1)) + (\sigma_2^2/n(H_2))}$$

and $z(\alpha)$ is the ordinate of the unit, one-sided, normal deviate. Note that $z(\alpha)$ is, in effect, a signal-to-noise ratio, where the signal in question is a selective advantage, and the noise is the collateral noise for the given schema competition.

For a given z, α can be found in standard tables, or approximated. For values of $|z| > 2$ (two standard deviations from the mean), one can use the Gaussian tail approximation:

$$\alpha = \frac{\exp\left(-z^2/2\right)}{\left(z(2\pi)^{1/2}\right)}.$$

† Technically, the central-limit theorem only applies to a random sample. Therefore, the assumption that the mean of observed, average fitness values is an unbiased sample of the average fitness values over all strings is only valid in the initial, random population, and perhaps in other populations early in the GA run. However, GA theory makes the assumption that selection is sufficiently slow to allow for good schema sampling.

For values of $|z| \leq 2$, one can use the sigmoidal approximation suggested by Valenzuela-Rendon (1989):

$$\alpha = \frac{1}{1 + \exp(-1.6z)}.$$

Given this calculation, one can match a desired maximum level of error in selection to a desired population size. This is accomplished by setting $n(H_1)$ and $n(H_2)$ such that the error probability is lowered below the desired level. In effect, raising either of the $n(H)$ values 'sharpens' (lowers the variance of) the associated normal distribution, thus reducing the convolution of the two distributions.

Goldberg *et al* (1992) suggest that if the largest value of $2^{O(H)}\sigma_m^2/|f(H_1) - f(H_2)|$ (where σ_m is the mean schemata variance, $(\sigma(H_1)^2 + \sigma(H_2)^2)/2)$) is known for competitive schemata of order $O(H)$, one can conservatively size the population by assuming the $n(H)$ values are the expected values for a random population of size μ. This gives the sizing formula:

$$\mu = 2z^2(\alpha)2^{O(H)}\frac{\sigma_m^2}{(f(H_1) - f(H_2))^2}.$$

Note that this formula can be extended to alphabets of cardinality greater than two.

The formula is a thorough compilation of the concepts of schema variance and its relationship to population sizing. However, it does present some difficulties. The values and ranges of $f(H)$ are not known beforehand for any schemata, although these values are implicitly estimated in the GA process. Moreover, the values of σ^2 are neither known nor estimated in the usual GA process. Despite these limitations, Goldberg *et al* (1991) suggest some useful rules of thumb for population sizing from this relationship.

For instance, consider problems with deception of order $k \ll \ell$. That is, all building blocks of order k or less have no deception. One could view such a function as the sum of $m = \ell/k$ subfunctions, f_i. Thus, the root-mean-squared variance of a subfunction is

$$\sigma_{rms}^2 = \frac{\sum_{i=1}^m \sigma_{f_i}^2}{m}.$$

An estimate of the variance of the average order-k schema is

$$\sigma_m^2 = (m - 1)\sigma_{rms}^2.$$

This gives a population of size

$$\mu = 2z^2(\alpha)2^k(m - 1)\frac{\sigma_{rms}^2}{(f(H_1) - f(H_2))^2}.$$

Note that the population size is $O(m) = O(\ell/k) = O(\ell)$ for problems of fixed, bounded deception order k.

This relationship suggests the rule of thumb that an adequate population size increases linearly with string length for problems of fixed, bounded deception. Moreover, it has some interesting implications for GA time complexity. Goldberg and Deb (1990) show that for typical selection schemes GAs converge in $O(\log \mu)$ or $O(\mu \log \mu)$ generations. This suggests that GAs can converge in $O(\ell \log \ell)$ generations, even when populations are sized to control selection errors.

One can construct another rule of thumb by considering the maximum variance in a GA fitness function, which is given by

$$\sigma_f^2 = \frac{(f_{\max} - f_{\min})^2}{4}$$

where $f_{\max}$ is the maximum fitness value, and $f_{\min}$ is the minimum fitness value for the function. One could use this value as a conservative estimate of the schema average fitness variance (the collateral noise), and size the population accordingly.

The population sizing formula has also suggested a method of dynamically adjusting population size. In a recent study (Smith 1993a, b, Smith and Smuda 1995), a modified GA is suggested that adaptively resizes the population based on the absolute expected selection loss, which is given by

$$L(H_1, H_2) = |f(H_1) - f(H_2)|\alpha(H_1, H_2)$$

where α is derived from the previous formula, and competitions of mates that estimate not only schema average finesses (as in the usual GA), but schema fitness variances as well. Note that the $L(H_1, H_2)$ measure considers not only the variance of a schema competition, but also its relative effect. Note that this is important in an adaptive sizing algorithm, since the previous population sizing formula does not consider the relative importance of schemata competitions. If two competing schemata have fitness values that are nearly equal, the overlap in the distributions will be great, thus suggesting a large population. However, if the fitness values of these schemata are nearly equal, their importance to the overall search may be minimal, thus precluding the need for a large population on their account.

Preliminary experiments with the adaptive population sizing technique have indicated its viability. They also suggest the possibility of other techniques that automatically and dynamically adjust population size in response to problem demands.

17.4   Final comments

This section has presented arguments for sizing GA populations. However, the concepts (maximizing computational leverage and ensuring accurate sampling) are general, and can be applied to other EC techniques. In different situations, either of these two concepts may determine the best population size. In many practical situations, it will be difficult to determine which concept dominates. Moreover, population size based on these concepts must be considered in the context of recombinative mixing, disruption, deception, population diversity, and selective pressure (Goldberg *et al* 1993). One must also consider the implementation details of a GA on parallel computers. Specifically, how does one distribute subpopulations on processors, and how does one exchange population members between processors? Some of these issues are considered in recent studies (Goldberg *et al* 1995). As EC methods advance, automatic balancing of these effects based on theoretical considerations is a prime concern.

References

De Jong K A 1975 An analysis of the behavior of a class of genetic adaptive systems *Dissertation Abstracts Int.* **36** 5140B (University Microfilms No 76–9381)

Goldberg D E 1989 Sizing populations for serial and parallel genetic algorithms *Proc. 3rd Int. Conf. on Genetic Algorithms (Fairfax, VA, June 1989)* ed J D Schaffer (San Mateo, CA: Morgan Kaufmann) pp 70–9

Goldberg D E and Deb K 1990 *A Comparative Analysis of Selection Schemes used in Genetic Algorithms* TCGA Report 90007, The University of Alabama, The Clearinghouse for Genetic Algorithms

Goldberg D E, Deb K and Clark J H 1991 *Genetic Algorithms, Noise, and the Sizing of Populations* IlliGAL Technical Report 91010, University of Illinois at Urbana-Champaign

——1992 Accounting for noise in the sizing of populations *Foundations of Genetic Algorithms 2* ed L D Whitley (San Mateo, CA: Morgan Kaufmann) pp 127–40

Goldberg D E, Deb K and Thierens D 1993 Toward a better understanding of mixing in genetic algorithms *J. Soc. Instrum. Control Eng.* **32** 10–16

Goldberg D E, Kargupta H, Horn J and Cantu-Paz E 1995 *Critical Deme Size for Serial and Parallel Genetic Algorithms* IlliGAL Technical Report 95002, University of Illinois at Urbana-Champaign

Goldberg D E and Rudnick M 1991 Genetic algorithms and the variance of fitness *Complex Syst.* **5** 265–78

Grefenstette J J 1986 Optimization of control parameters for genetic algorithms *IEEE Trans. Syst. Man Cybernet.* **SMC-16** 122–8

Krishnakumar K 1989 Microgenetic algorithms for stationary and non-stationary function optimization *SPIE Proc. on Intelligent Control and Adaptive Systems* vol 1196 (Bellingham, WA: SPIE) pp 289–96

Schaffer J D, Caruana R A, Eshelman L J and Das R 1989 A study of control parameters affecting online performance of genetic algorithms for function optimization *Proc. 3rd Int. Conf. on Genetic Algorithms (Fairfax, VA, June 1989)* (San Mateo, CA: Morgan Kaufmann) pp 51–60

Smith R E 1993a *Adaptively Resizing Populations: an Algorithm and Analysis* TCGA Report 93001, University of Alabama

——1993b Adaptively resizing populations: an algorithm and analysis *Proc. 5th Int. Conf. on Genetic Algorithms (Urbana-Champaign, IL, July 1993)* (San Mateo, CA: Morgan Kaufmann) p 653

Smith R E and Smuda E 1995 Adaptively resizing populations: algorithm, analysis, and first results *Complex Syst.* **9** 47–72

Valenzuela-Rendon M 1989 *Two Analysis Tools to Describe the Operation of Classifier Systems* TCGA Report 89005, The University of Alabama, The Clearinghouse for Genetic Algorithms

18

Mutation parameters

Thomas Bäck

18.1 Introduction

The basic distinction between the concept of handling mutation in evolution strategies and evolutionary programming as opposed to genetic algorithms can be clarified as follows: evolution *strategies* and evolutionary programming evolve their set of mutation parameters ($n_\sigma \in \{1, \ldots, n\}$ variances and $n_\alpha \in \{0, \ldots, (n - n_\sigma/2)(n_\sigma - 1)\}$ covariances of the generalized, n-dimensional normal distribution) on-line during the search by applying the search operator(s) mutation (and recombination, in the case of evolution strategies) to the strategy parameters as well. This principle facilitates the self-adaptation (Chapter 21) of strategy parameters and shifts the parameter setting issue to the more robust level of the learning rates; that is, the parameters that control the speed of the adaptation of strategy parameters. Section 18.2 briefly discusses the presently used heuristics (which are based on some theoretical ground) for setting these learning rates on the meta-level of strategy parameter modifications.

In contrast to associating a (potentially large) number of mutation parameters with each single individual and self-adapting these parameters on-line, genetic algorithms and evolutionary heuristics derived from genetic algorithms usually provide only one mutation rate p_m for the complete population. This mutation rate is set to a fixed value, and it is not modified or self-adapted during evolution. A variety of values were proposed for setting p_m, and a summary of these results (which are obtained from experimental investigations) is given in Section 18.3. In addition to constant settings of the mutation rate, some experiments with a mutation rate varying over the generation number are also provided in the literature, including efforts to calculate the optimal schedule of the mutation rate for simple objective functions and to derive some general heuristics from these investigations. Furthermore, the variation of the mutation rate might be probabilistic rather than deterministic, and p_m might also vary over bit representation in the case of binary representation of individuals. These mutation heuristics are discussed in Section 18.3.

18.2 Mutation parameters for self-adaptation

In evolution strategies, the mutation of standard deviations σ_i ($i \in \{1, \ldots, n_\sigma\}$) according to the description given by Bäck and Schwefel (1993); that is,

$$\sigma_i' = \sigma_i \exp(\tau' N(0, 1) + \tau N_i(0, 1)) \tag{18.1}$$

is controlled by two meta-parameters or learning rates τ and τ'. Schwefel (1977, p 167–8) suggests setting these parameters according to

$$\tau' = \frac{K}{(2n)^{1/2}} \quad \text{and} \quad \tau = \frac{K}{[2(2n)^{1/2}]^{1/2}} \tag{18.2}$$

and recently Schwefel and Rudolph (1995) generalized this rule by setting

$$\tau' = \frac{K\delta}{(2n)^{1/2}} \quad \text{and} \quad \tau = \frac{K(1-\delta)}{[2n/(n_\sigma)^{1/2}]^{1/2}} \tag{18.3}$$

where K denotes the (in general unknown) normalized convergence velocity of the algorithm. Although the convergence velocity K cannot be known for arbitrary problems, the parameters τ and τ' are very robust against settings deviating from the optimal value (a variation within one order of magnitude normally causes only a minor loss of efficiency). Consequently, a setting of $K = 1$ is a useful initial recommendation.

Experiments varying the weighting factor δ have not been performed so far, such that the default value $\delta = 1/2$ should be used for first experiments with an evolution strategy.

For the mutation of rotation angles α_j in evolution strategies with correlated mutations, which is performed according to the rule

$$\alpha_j' = \alpha_j + \beta N_j(0, 1) \tag{18.4}$$

a value of $\beta = 0.0853$ (corresponding to 5°) is recommended on the basis of experimental results.

For the simple self-adaptation case $n_\sigma = 1$, where only one standard deviation is learned per individual, equation (18.1) simplifies to $\sigma' = \sigma \exp(\tau_0 N(0, 1))$ with a setting of $\tau_0 = K/n^{1/2}$. Alternatively, Rechenberg (1994) favors a so-called mutational step size control which modifies σ according to the even simpler rule $\sigma' = \sigma u$, where $u \sim U(\{1, \alpha, 1/\alpha\})$ is a uniform random value attaining one of the values 1, α and $1/\alpha$. This is motivated by the idea of trying larger, smaller, and constant standard deviations in the next generation, each of these with one-third of the individuals. Rechenberg (1994, p 48) recommends a value of $\alpha = 1.3$ for the learning rate. An experimental comparison of both self-adaptation rules for $n_\sigma = 1$ has not been performed so far.

In the case of evolutionary programming (EP), Fogel (1992) originally proposed an additive, normally distributed mutation of variances for self-adaptation in meta-EP, but subsequently substituted the same logarithmic-normally distributed modification as used in evolution strategies (Saravanan and Fogel 1994a, 1994b). Consequently, the parameter setting rules for τ' and τ also apply to evolutionary programming.

18.3 Mutation parameters for direct schedules

Holland (1975) introduced the mutation operator of genetic algorithms as a 'background operator' that changes bits of the individuals only occasionally, with a rather small mutation probability $p_m \in [0, 1]$ per bit. Common settings of the mutation probability are summarized in table 18.1.

Table 18.1. Commonly used constant settings of the mutation rate p_m in genetic algorithms.

p_m	Reference
0.001	De Jong (1975, pp 67–71)
0.01	Grefenstette (1986)
0.005–0.01	Schaffer *et al* (1989)

These settings were all obtained by experimental investigations, including a meta-level optimization experiment performed by Grefenstette (1986), where the space of parameter values of a genetic algorithm was searched by another genetic algorithm. Mutation rates within the range of values summarized in table 18.1 are still widely used in applications of canonical (i.e. using binary representation) genetic algorithms, because these settings are consistent with Holland's proposal for mutation as a background operator and Goldberg's recommendation to invert on the order of one per thousand bits by mutation (Goldberg 1989, p 14). Although it is correct that base pair mutations of *Eschericia coli* bacteria occur with a similar frequency (Futuyma 1990, pp 82–3), it is important to bear in mind that this reflects mutation rates in a relatively late stage of evolution and in only one specific example of natural evolution, which may or may not be relevant to genetic algorithms. Early stages in the history of evolution on earth, however, were characterized by much larger mutation rates (see Ebeling *et al* 1990, ch 8).

Taking this into account, some authors proposed varying the mutation rate in genetic algorithms over the number of generations according to some specific, typically decreasing schedule—which is usually deterministic, but might also be probabilistic. Fogarty (1989) performed experiments comparing, for binary strings of length $\ell = 70$, the following schedules:

(i)   a constant mutation rate $p_m = 0.01$

(ii) a mutation rate

$$p_m(t) = \frac{1}{240} + \frac{0.11375}{2^t} \qquad (18.5)$$

that is, a schedule where the mutation rate decreases exponentially over time

(iii) a mutation rate varying over bit representations but not over generations, setting $p_m(i)$ for bit number $i \in \{1, \ldots, \ell\}$ ($i = 1$ indicates the least significant bit) to a value

$$p_m(i) = \frac{0.3528}{2^{i-1}}. \tag{18.6}$$

(iv) a combination of both according to

$$p_m(i, t) = \frac{28}{1905 \times 2^{i-1}} + \frac{0.4026}{2^{t+i-1}}. \tag{18.7}$$

The graphs of these schedules (ii)–(iv) are shown in figures 18.1 and 18.2 to give an impression of their general form. It is worth noting that the schedule according to (ii) decreases quickly within less than 10 generations to the baseline value.

For a specific application problem, Fogarty arrived at the conclusion that varying mutation rates over generations and/or across integer representation significantly improves the on-line performance of a genetic algorithm, if evolution is started with a population of all zero bits. Although this result was obtained for a specific experimental setup, Fogarty's investigations serve as an important starting point for other studies about varying mutation rates.

Hesser and Männer (1991, 1992) succeeded in deriving a general expression for a time-varying mutation rate of the form

$$p_m(t) = \left(\frac{c_1}{c_2}\right)^{1/2} \frac{\exp(-c_3 t / 2)}{\mu (\ell)^{1/2}} \tag{18.8}$$

which favors an exponentially decreasing mutation rate and seems to confirm Fogarty's findings. Furthermore, equation (18.8) also contains the population size μ as well as the string length ℓ as additional parameters which are relevant for the optimal mutation rate, and the dependence on these parameters shows some correspondence with the empirical finding $p_m \approx 1.75/(\mu(\ell)^{1/2})$ obtained by Schaffer et al (1989) by curve fitting of their experimental data. Unfortunately, the constants c_i are generally unknown and can be estimated only for simple cases from heuristic arguments, such that equation (18.8) does not offer a generally useful rule for setting p_m.

Recently, some results concerning optimal schedules of the mutation rate in the cases of simple objective functions and simplified genetic algorithms were presented by Mühlenbein (1992), Bäck (1992a, 1993) and Yanagiya (1993). This work is based on the idea of finding a schedule that maximizes the convergence velocity or minimizes the *absorption time* of the algorithm (i.e. the number of iterations until the optimum is found). To facilitate the theoretical analysis, these authors work with the concept of a $(1 + \lambda)$- or $(1, \lambda)$-genetic

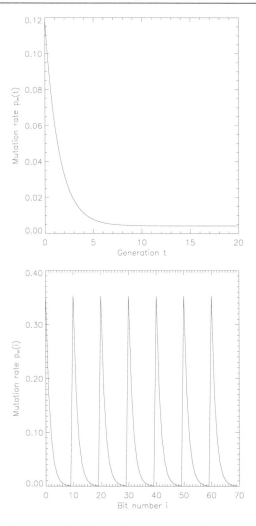

Figure 18.1. Mutation rate schedule varying over generation number according to the
description given in Fogarty's schedule (ii) (top) and over bit representation according
to the description given in his schedule (iii) (bottom).

algorithm (most often, a $(1 + 1)$-algorithm is considered). Such an algorithm is
characterized by a single parent individual, generating λ offspring individuals
by mutation. In the case of plus-selection, the best of parent and offspring is
selected for the next generation, while the best of offspring only is selected in
the case of comma-selection. For the simple 'counting ones' objective function
$f(b_1 \ldots b_\ell) = \sum_{i=1}^{\ell}$, Bäck (1992a, 1993) demonstrated that a mutation rate
schedule starting with $p_m(f(\boldsymbol{b}) = \ell/2) = 1/2$ and decreasing exponentially
towards $1/\ell$ as $f(\boldsymbol{b})$ approaches ℓ is optimal, and he presented an approximation

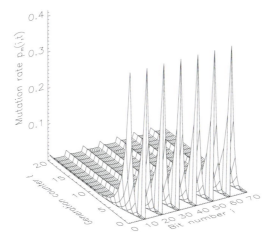

Figure 18.2. Mutation rate schedule varying over both bit representation and generation number according to the description given in Fogarty's schedule (iv).

for a $(1 + 1)$-genetic algorithm where

$$p_{\mathrm{m}}(f(b)) \approx \frac{1}{2(f(b) + 1) - \ell} \qquad (18.9)$$

defines the mutation rate as a function of $f(b)$ at generation t. As a more useful result, however, both Mühlenbein (1992) and Bäck (1993) concluded that a constant mutation rate

$$p_{\mathrm{m}} = 1/\ell \qquad (18.10)$$

is almost optimal for a $(1 + 1)$-genetic algorithm applied to this problem and can serve as a reasonable heuristic rule for any kind of objective function, because it is impossible to derive analytical results for complex functions. This result, however, was already provided by Bremermann *et al* (1966), who used the same approximation method that Mühlenbein used 26 years later. Yanagiya (1993) and Bäck (1993) also presented optimal mutation rate schedules for more complicated objective functions such as quasi-Hamming-distance functions (the objective function value depends strongly on the Hamming distance from the global optimum), a knapsack problem, and decoding functions mapping binary strings to integers. The resulting optimal mutation rate schedules are often quite irregular and utilize surprisingly large mutation rates, but it seems impossible to draw a general conclusion from these results. Certainly, the above-mentioned rule, $p_{\mathrm{m}} = 1/\ell$, is always a good starting point because it will not perform worse than any smaller mutation rate setting. As the number of offspring individuals increases, however, the optimal mutation rate as well as the associated convergence velocity increase also, but currently no useful analytical results are known for the dependence of the optimal mutation rate on offspring population size (see Bäck 1996, chapter 6).

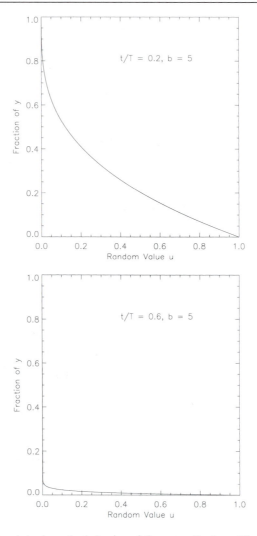

Figure 18.3. The plots show the behavior of the normalized modification $\Delta x_i(t, y)/y$ as a function of the random number u according to operator (18.11). In both cases, the default value $b = 5$ was chosen. The upper plot ($t/T = 0.2$) demonstrates that the modifications occurring in the early stages of a run are quite large, while later on only small modifications are possible, as shown in the lower plot ($t/T = 0.6$).

In addition to deterministic schedules, nondeterministic schedules for controlling the 'amount' (in the sense of continuous step-sizes as well as the probability to invert bits) of mutation are also known in the literature and applied in the context of evolutionary algorithms. Michalewicz (1994) introduced a step-size control mechanism for real-valued vectors, which decreases the amount

Δx_i of the modification of an object variable x_i over the number of generations according to

$$\Delta x_i(t, y) = y \left(1 - u^{(1-t/T)^b}\right) \tag{18.11}$$

where $u \sim U([0, 1])$ is a uniform random number, T is the maximum generation number, y is the maximum value of the modification Δx_i, and b is a system parameter determining the degree of dependency on t. Michalewicz (1994, p 101) proposes a value of $b = 5$. In contrast to evolution strategies or evolutionary programming, only a single, randomly chosen (according to a uniform distribution $U(\{1, \ldots, n\})$) object variable is modified when mutation is applied; that is, $m(x_1, \ldots, x_n) = (x_1, \ldots, x_{i-1}, x_i', x_{i+1}, \ldots, x_n)$.

Using equation (18.11), the modification of the selected object variable x_i is performed according to

$$x_i' = \begin{cases} x_i + \Delta x_i(t, \overline{x}_i - x_i) & \text{if } u' = 1 \\ x_i - \Delta x_i(t, x_i - \underline{x}_i) & \text{if } u' = 0 \end{cases} \tag{18.12}$$

where $u' \sim U(\{0, 1\})$ is a uniform random digit, and $\overline{x}_i$, $\underline{x}_i$ denote upper and lower domain bounds of x_i. The plots in figure 18.3 show the normalized modification $\Delta x_i(t, y)/y$ as a function of the random variable u for $b = 5$ at two different time stages of a run. Clearly, the possible modification of x_i decreases (quickly, for this value of b) as the generation counter increases.

To facilitate a comparison with a binary representation of object variables, Michalewicz (1994) also modeled this mutation operator for the binary space $\{0, 1\}^\ell$. Again, the object variable x_i, which should be modified by mutation, is randomly chosen from the n object variables. For the binary representation $(b_1, \ldots, b_{l_x})$ of x_i (with a length of $l_x = 30$ bits per object variable), the mutation operator inverts the value of the bit $b_{\Delta'(t, l_x)}$, where the bit position $\Delta'(t, l_x)$ is defined by

$$\Delta'(t, l_x) = 1 + \begin{cases} \lceil \Delta(t, l_x) \rceil & \text{if } u' = 1 \\ \lfloor \Delta(t, l_x) \rfloor & \text{if } u' = 0 \end{cases} \tag{18.13}$$

and the parameter $b = 1.5$ is chosen to achieve similar behavior as in (18.11) (here, we have to add a value of one because, in contrast to Michalewicz, we consider the least significant bit being indexed by one rather than zero). The effect of this operator is to concentrate mutation on the less significant bits as the generation counter t increases, therefore causing smaller and smaller modifications on the level of decoded object variables. Consequently, this probabilistic mutation rate schedule has some similarity to Fogarty's settings (iii) and (iv), varying the mutation rate both over bit representation and over time.

A comparison of both operators clearly demonstrated a better performance for the operator (18.11) designed for real-valued object variables (Michalewicz 1994, p 102).

18.4 Summary

Optimal setting of the mutation rate or mutation step size(s) in evolutionary algorithms is not an easy task. Use of the self-adaptation technique simplifies the problem by switching to meta-parameters which determine the speed of step size adaptation rather than the step sizes themselves. The new meta-parameters are generally robust and their default settings work well in many cases, though they do not guarantee the fastest adaptation in the general case.

Direct control mechanisms of mutation rate or step size are typically applied in genetic algorithms and their derivates. Usually, a constant mutation rate is utilized, although it is well known that no generally valid best mutation rate exists. It is known, however, that $p_m = 1/\ell$ is to be preferred over smaller values and can serve as a general guideline if nothing is known about the objective function.

Very little is known concerning mutation rate schedules varying over time or bit representations—except the theoretical results indicating the superiority of schedules depending on the distance to the optimum. In case of the 'counting ones' problem, the decrease of p_m over time is exponential, which seems to be a promising choice also for more difficult objective functions. The alternative way of self-adapting the mutation rate in genetic algorithms, which certainly opens an interesting and promising path towards solving this parameter setting problem, has not yet been exploited in depth (except in some preliminary work by Bäck (1992b)).

References

Bäck T 1992a The interaction of mutation rate, selection, and self-adaptation within a genetic algorithm *Proc. 2nd Int. Conf. on Parallel Problem Solving from Nature (Brussels, 1992)* ed R Männer and B Manderick (Amsterdam: Elsevier) pp 85–94

——1992b Self-adaptation in genetic algorithms *Proc. 1st Eur. Conf. on Artificial Life* ed F J Varela and P Bourgine (Cambridge, MA: MIT Press) pp 263–71

——1993 Optimal mutation rates in genetic search *Proc. 5th Int. Conf. on Genetic Algorithms (Urbana-Champaign, IL, July 1993)* ed S Forrest (San Mateo, CA: Morgan Kaufmann) pp 2–8

——1996 *Evolutionary Algorithms in Theory and Practice* (New York: Oxford University Press)

Bäck T and Schwefel H-P 1993 An overview of evolutionary algorithms for parameter optimization *Evolut. Comput.* **1** 1–23

Bremermann H J and Rogson M and Salaff S 1966 Global properties of evolution processes *Natural Automata and Useful Simulations* ed H H Pattec *et al* (Washington, DC: Spartan Books) pp 3–41

De Jong K A 1975 *An analysis of the behavior of a class of genetic adaptive systems* PhD Thesis, University of Michigan

Ebeling W and Engel A and Feistel R 1990 *Physik der Evolutionsprozesse* (Berlin: Akademie-Verlag)

Fogarty T C 1989 Varying the probability of mutation in the genetic algorithm *Proc. 3rd Int. Conf. on Genetic Algorithms (Fairfax, VA, June 1989)* ed J D Schaffer (San Mateo, CA: Morgan Kaufmann) pp 104–9

Fogel D B 1992 *Evolving Artificial Intelligence* PhD Thesis, University of California, San Diego

Futuyma D J 1990 *Evolutionsbiologie* (Basel: Birkhäuser)

Goldberg D E 1989 *Genetic algorithms in search, optimization and machine learning* (Reading, MA: Addison-Wesley)

Grefenstette J J 1986 Optimization of control parameters for genetic algorithms *IEEE Transactions on Systems, Man and Cybernetics* **SMC–16** 122–8

Hesser J and Männer R 1991 Towards an optimal mutation probability in genetic algorithms *Proc. 1st Workshop on Parallel Problem Solving from Nature (Dortmund, 1990) (Lecture Notes in Computer Science 496)* ed H-P Schwefel and R Männer (Berlin: Springer) pp 23–32

——1992 Investigation of the m-heuristic for optimal mutation probabilities *Proc. 2nd Int. Conf. on Parallel Problem Solving from Nature (Brussels, 1992)* ed R Männer and B Manderick (Amsterdam: Elsevier) pp 115–24

Holland J H 1975 *Adaptation in Natural and Artificial Systems* (Ann Arbor, MI: University of Michigan Press)

Michalewicz Z 1994 *Genetic Algorithms + Data Structures = Evolution Programs* (Berlin: Springer)

Mühlenbein H 1992 How genetic algorithms really work: I. Mutation and hillclimbing *Proc. 2nd Int. Conf. on Parallel Problem Solving from Nature (Brussels, 1992)* ed R Männer and B Manderick (Amsterdam: Elsevier) pp 15–25

Rechenberg I 1994 *Evolutionsstrategie '94* Werkstatt Bionik und Evolutionstechnik, vol 1 (Stuttgart: Frommann–Holzboog)

Saravanan N and Fogel D B 1994a Learning of strategy parameters in evolutionary programming: an empirical study *Proc. 3rd Ann. Conf. on Evolutionary Programming (San Diego, CA, February 1994)* ed A V Sebald and L J Fogel (Singapore: World Scientific) pp 269–80

——1994b Evolving neurocontrollers using evolutionary programming *Proc. 1st IEEE Conf. on Evolutionary Computation (Orlando, FL, June 1994)* (Piscataway, NJ: IEEE) pp 217–22

Schaffer J D, Caruana R A, Eshelman L J and Das R 1989 A study of control parameters affecting online performance of genetic algorithms for function optimization *Proc. 3rd Int. Conf. on Genetic Algorithms (Fairfax, VA, June 1989)* ed J D Schaffer (San Mateo, CA: Morgan Kaufmann) pp 51–60

Schwefel H-P 1977 *Numerische Optimierung von Computer-Modellen mittels der Evolutionsstrategie (Interdisciplinary Systems Research 26)* (Basel: Birkhäuser)

Schwefel H-P and Rudolph G 1995 Contemporary evolution strategies *Advances in Artificial Life :(Proc. 3rd Int. Conf. on Artificial Life) (Lecture Notes in Artificial Intelligence 929)* ed F Morán et al (Berlin: Springer) pp 893–907

Yanagiya M 1993 A simple mutation-dependent genetic algorithm *Proc. 5th Int. Conf. on Genetic Algorithms (Urbana-Champaign, IL, July 1993)* ed S Forrest (San Mateo, CA: Morgan Kaufmann) p 695

19

Recombination parameters

William M Spears

19.1 General background

Although Holland (1975) was not the first to suggest recombination in an evolutionary algorithm (EA) (see e.g. Fraser 1957, Fogel *et al* 1966), he was the first to place theoretical emphasis on this operator. This emphasis stemmed from his work in adaptive systems, which resulted in the field of genetic algorithms (GAs) and genetic programming. According to Holland, an adaptive system must persistently test and incorporate structural properties associated with better performance. The object, of course, is to find *new* structures which have a high probability of improving performance significantly.

Holland concentrated on schemata, which provide a basis for associating combinations of attributes with potential for improving current performance. To see this, let us consider the schema AC##, defined over a fixed length chromosome of four genes, where each gene can take on one of three alleles {A, B, C}. If # is defined to be a 'don't care' (i.e. wildcard) symbol, the schema AC## represents all chromosomes that have an A for their first allele and a C for their second. Since each of the # symbols can be filled in with any one of the three alleles, this schema represents $3^2 = 9$ chromosomes.

Suppose every chromosome has a well-defined fitness value (also called *utility* or *payoff*). Now suppose there is a population of P individuals, p of which are members of the above schema. The *observed average fitness* of that schema is the average fitness of these p individuals in that schema. It is important to note that these individuals will also be members of other schemata, thus the population of P individuals contains instances of a large number of schemata (all of which have some observed fitness). Holland (1975) stated that a good heuristic is to generate new instances of these schemata whose observed fitness is higher than the average fitness of the whole population, since instances of these schemata are likely to exhibit superior performance.

Suppose the schema AC## does in fact have a high observed fitness. The heuristic states that new samples (instances) of that schema should be generated. Selection (reproduction) does not produce new samples—but recombination can.

The key aspect of recombination is that if one recombines two individuals that start with AC, their offspring must also start with AC. Thus one can retain what appears to be a promising building block (AC##), yet continue to test that building block in new contexts.

As stated earlier, recombination can be implemented in many different ways. Some forms of recombination are more appropriate for certain problems than are others. According to Booker (1992) it is thus useful to characterize the biases in recombination operators, recognize when these biases are correct or incorrect (for a given problem or problem class), and recover from incorrect biases when possible. Sections 19.2 and 19.3 summarize much of the work that has gone into characterizing the biases of various recombination operators. Historically, most of the earlier recombination operators were designed to work on low-level 'universal' representations, such as the fixed-length low-cardinality representation shown above. In fact, most early GAs used just simple bitstring representations. A whole suite of recombination operators evolved from that level of representation. Section 19.2 focuses on such 'bit-level' recombination operators. Recent work has focused more on problem-class-specific representations, with recombination operators designed primarily for these representations. Section 19.3 focuses on the more problem-class-specific recombination operators.

Section 19.4 summarizes some of the mechanisms for recognizing when biases are correct or incorrect, and recovering from incorrect biases when possible. The conclusion outlines some design principles that are useful when creating new recombination operators. Ideally, one would like firm and precise practical rules for choosing what form and rate of recombination to use on a particular problem; however, such rules have been difficult to formulate. Thus this section concentrates more on heuristics and design principles that have often proved useful.

19.2   Genotypic-level recombination

19.2.1   Theory

Holland (1975) provided one of the earliest analyses of a recombination operator, called one-point recombination. Suppose there are two parents: ABCD and AABC. Randomly select one point at which to separate (*cut*) both parents. For example, suppose they are cut in the middle (AB|CD and AA|BC). The offspring are created by swapping the tail (or head) portions to yield ABBC and AACD. Holland analyzed one-point recombination by examining the probability that various schemata will be disrupted when undergoing recombination. For example, consider the two schemata AA## and A##A. Each schema can be disrupted only if the *cut point* falls between its two As. However, this is much more likely to occur with the latter schema (A##A) than the former (AA##). In fact, the probability of disrupting either schema is proportional to the distance between the As. Thus, one-point recombination has the bias that it is much more likely to disrupt 'long' schemata than 'short' schemata, where the length of a schema is the distance between the first and the last *defining position* (a nonwildcard).

De Jong (1975) extended this analysis to include so-called *n-point* recombination. In *n*-point recombination *n* cut points are randomly selected and the genetic material between cut points is swapped. For example, with two-point recombination, suppose the two parents ABCD and AABC are cut as follows: A|BC|D and A|AB|C. Then the two offspring are AABD and ABCC. De Jong noted that two-point (or *n*-point where *n* is even) recombination is less likely to disrupt 'long' schemata than one-point (or *n*-point where *n* is odd) recombination.

Syswerda (1989) introduced a new form of recombination called *uniform* recombination. Uniform recombination does not use *cut points* but instead creates offspring by deciding, for each allele of one parent, whether to swap that allele with the corresponding allele in the other parent. That decision is made using a coin flip (i.e. the swap is made 50% of the time). Syswerda compared the probability of schema disruption for one-point, two-point, and uniform recombination. Interestingly, while uniform recombination is somewhat more disruptive of schemata than one-point and two-point, it does not have a length bias (i.e. the length of a schema does not affect the probability of disruption). Also, Syswerda showed that the more disruptive nature of uniform recombination can be viewed in another way—it is more likely to construct instances of new schemata than one-point and two-point recombination.

De Jong and Spears (1992) verified Syswerda's results and introduced a parameterized version of uniform recombination (where the probability of swapping could be other than 50%). Lowering the swap probability of uniform recombination allows one to lower disruption as much as desired, while maintaining the lack of length bias. Finally, De Jong and Spears characterized recombination in terms of two other measures: *productivity* and *exploration power*. The productivity of a recombination operator is the probability that it will generate offspring that are different from their parents. More disruptive recombination operators are more productive (and vice versa). An operator is more *explorative* if it can reach a larger number of points in the space with one application of the operator. Uniform recombination is the most explorative of the recombination operators since, if the Hamming distance between two parents is h (i.e. h loci have different alleles), uniform recombination can reach any of 2^h points in one application of the operator. Moon and Bui (1994) independently performed a similar analysis. Although mathematically equivalent, this analysis emphasized 'clusters' of defining positions within schemata, as opposed to lengths.

Eshelman *et al* (1989) considered other characterizations of recombination bias. They introduced two biases, the *positional* and *distributional bias*. A recombination operator has positional bias to the extent that the creation of any new schema by recombining existing schemata is dependent upon the location of the alleles in the chromosome. This is similar to the length bias introduced above. A recombination operator has distributional bias to the extent that the amount of material that is expected to be exchanged is distributed around some value or values ranging from 1 to $L - 1$ alleles (where the chromosome is composed of L genes), as opposed to being uniformly distributed. For example, one-point recombination has high positional and no distributional bias, while two-point recombination has slightly lower positional bias and still no distributional bias. Uniform recombination has no positional bias but high distributional bias because the amount of material exchanged is binomially distributed. Later, Eshelman and Schaffer (1994) refined their earlier study and introduced *recombinative* bias, which is related to their older distributional bias. They also introduced *schema* bias, which is a generalization of their older positional bias.

Booker (1992) tied the earlier work together by characterizing recombination operators via their recombination distributions, which describe the probability of all possible recombination events. The recombination distributions were used to rederive the disruption analysis of De Jong and Spears (1992) for n-point and parameterized uniform recombination, as well as to calculate precise values for the distributional and positional biases of recombination. This reformulation allowed Booker to detect a symmetry in the positional bias of n-point recombination around $n = L/2$, which corrected a prediction made by Eshelman *et al* (1989) that positional bias would continue to increase as n increases.

19.2.2 Heuristics

The sampling arguments and the characterization of biases that has just been presented have motivated a number of heuristics for how to use recombination and how to choose which recombination to use.

Booker (1982) considered implementations of recombination from the perspective of trying to improve overall performance. The motivation was that allele loss from the population could hurt the sampling of coadapted sets of alleles (schemata). In the earliest implementations of GAs one offspring of a recombination event would be thrown away. This was a source of allele loss, since, instead of transmitting all alleles from both parents to the next generation, only a subset was transmitted. The hypothesis was that allele loss rates would be greatly decreased by saving both offspring. That hypothesis was confirmed empirically. There was also some improvement in on-line (average fitness of all samples) and off-line (average fitness of the best samples) performance on the De Jong (1975) test suite, although the off-line improvement was negligible.

Booker (1987) also pointed out that, due to allele loss, recombination is less likely to produce children different from their parents as a population evolves. This effectively reduces the sampling of new schemata. To counteract this

Booker suggested a more explorative version of recombination, termed *reduced surrogate* recombination, that concentrates on those portions of a chromosome in which the alleles of two parents are not the same. This ensures that a new sample is created. For example, suppose two parents are ABCD and ADBC. Then if one uses one-point recombination, and the cut point occurs immediately after the A, the two offspring would be identical to the parents. Reduced surrogate recombination would ensure that the cut point was further to the right.

It has been much more difficult to come up with heuristics for choosing which recombination operator to use in a given situation. Syswerda (1989), however, noted that one nice aspect of uniform recombination is that, due to its lack of length bias, it is not affected by the presence of irrelevant alleles in a representation. Nor is it affected by the position of the relevant alleles on the chromosome. Thus, for those problems where little information is available concerning the relevance of alleles or the length of building blocks, uniform recombination is a useful default.

De Jong and Spears (1990) tempered this view somewhat, by including interactions with population size (Chapter 17). Their heuristic was that disruption is most useful when the population size is small or when the population is almost homogeneous. They argued that more disruptive recombination operators (such as 0.5 uniform recombination, or n-point recombination where $n > 2$) should be used when the population size is small relative to the problem size, and less disruptive recombinations operators (such as two-point, or uniform recombination with a swap probability less than 50%) should be used when the population size is large relative to the problem size. De Jong and Spears demonstrated this with a series of experiments in which the population size was varied.

Schaffer *et al* (1989) made a similar observation. They concentrated on the *recombination rate*, which is the percentage of the population to undergo recombination. They observed that high recombination rates are best with small populations, a broad range of recombination rates are tolerated at medium population sizes, and only low recombination rates are suggested for large population sizes.

Finally, Eshelman and Schaffer (1994) have attempted to match the biases of recombination operators with various problem classes and GA behavior. They concluded with two heuristics. The first was that high schema bias can lead to hitchhiking, where the EA exploits spurious correlations between schemata that contribute to performance and other schemata that do not. They recommended using a high-recombinative-bias and low-schema-bias recombination to combat premature convergence (i.e. loss of genetic diversity) due to hitchhiking. The second heuristic was that high recombinative bias can be detrimental in trap problems.

19.3 Phenotypic-level recombination

19.3.1 Theory

Thus far the focus has been on fixed-length representations in which each gene can take on one of a discrete set of alleles (values). Schemata were then defined, each of which represent a set of chromosomes (the chromosomes that match alleles on the defining positions of the schema). However, there are problems that do not match these representations well. In these cases new representations, recombination operators, and theories must be developed.

For example, a common task is the optimization of some real-valued function of real values. Of course, it is possible to code these real values as bitstrings in which the degree of granularity is set by choosing the appropriate number of bits. At this point conventional schema theory may be applied. However, there are difficulties that arise using this representation. One is the presence of *Hamming cliffs*, in which large changes in the binary encoding are required to make small changes to the real values. The use of Gray codes does not totally remove this difficulty. Standard recombination operators also can have the effect of producing offspring far removed (in the real-valued sense) from their parents (Schwefel 1995). An alternative representation is to simply use chromosomes that are real-valued vectors. In this case a more natural recombination operator averages (blends) values within two parent vectors to create an offspring vector. This has the nice property of creating offspring that are near the parents. (See the work of Davis (1991), Wright (1991), Eshelman, and Schaffer (1992), Schwefel (1995), Peck and Dhawan (1995), Beyer (1995), and Arabas *et al* (1995) for other recombination operators that are useful for real-valued vectors. One recombination operator of note is discrete recombination, which is the analog of uniform recombination on real-valued variables.)

Recently, three theoretical studies analyzed the effect of recombination using real-valued representations. Peck and Dhawan (1995) showed how various properties of recombination operators can influence the ability of the EA to converge to the global optima. Beyer (1995) concluded that an important role of recombination in this context is genetic repair, diminishing the influence of harmful mutations. Eshelman and Schaffer (1992) analyzed this particular representation by restricting the parameters to be integer ranges. Their *interval schemata* represent subranges. For example, if a parameter has range [0, 2], it has the following interval schemata: [0], [1], [2], [0, 1], [1, 2], and [0, 2]. The chromosomes (1 1 2) and (2 1 2) are instances of the interval schema ([1, 2] [1] [2]). Long interval schemata are more general and correspond roughly to traditional schemata that contain a large number of #s. Eshelman and Schaffer used *interval schemata* to help them predict the failure modes of various real-valued recombination operators.

Another class of important tasks involves permutation or ordering problems, in which the ordering of alleles on the chromosome is of primary importance. A large number of recombination operators have been suggested for these tasks, including partially mapped recombination (Goldberg and Lingle 1985), order recombination (Davis 1985), cycle recombination (Oliver *et al* 1987), edge recombination (Starkweather *et al* 1991), and the group recombination of Falkenauer (1994). Which operators work best depend on the objective function.

A classic permutation problem is the traveling salesman problem (TSP). Consider a TSP of four cities {A, B, C, D}. It is important to note that there are only 4! possible chromosomes, as opposed to 4^4 (e.g. the chromosome AABC is not valid). Also, note that schema ##BC does not have the same meaning as before, since the alleles that can be used to fill in the #s now depend on the alleles in the defining positions (e.g. a B or a C can not be used in this case). This led Goldberg and Lingle (1985) to define o-schemata, in which the 'don't cares' are denoted with !s. An example of an o-schema is !!BC, which defines the subset of all orderings that have BC in the third and fourth positions. For this example there are only two possible orderings, ADBC and DABC. Goldberg considered these to be *absolute* o-schemata, since the absolute position of the alleles is of importance. An alternative would be to stress the relative positions of the alleles. In this case what is important about !!BC is that B and C are adjacent—!!BC, !BC!, and BC!! are now equivalent o-schemata. One nice consequence of the invention of o-schemata is that a theory similar to that of the more standard schema theory can be developed. The interested reader is encouraged to see the article by Oliver *et al* (1987) for a nice example of this, in which various recombination operators are compared via an o-schema analysis.

There has also been some work on recombination for finite-state machines (Fogel *et al* 1966), variable-length chromosomes (Smith 1980, De Jong *et al* 1993), chromosomes that are LISP expressions (Fujiki and Dickinson 1987, Koza 1994, Rosca 1995), chromosomes that represent strategies (i.e. rule sets) (Grefenstette *et al* 1990), and recombination for multidimensional chromosomes (Kahng and Moon 1995). Some theory has recently been developed in these areas. For example, Bui and Moon (1995) developed some theory on multidimensional recombination. Also, Radcliffe (1991) generalized the notion of schemata to sets he refers to as formae.

19.3.2 *Heuristics*

Due to the prevalence of the traditional bitstring representation in GAs, less work has concentrated on recombination operators for higher-level representations, and there are far fewer heuristics. The most important heuristic is that recombination must identify and combine meaningful building blocks of chromosomal material. Put another way, 'recombination must take into account the interaction among the genes when generating new instances' (Eshelman and Schaffer 1992). The conclusion of this section provides some guidance in how to achieve this.

19.4 Control of recombination parameters

As can be seen from the earlier discussion, there is very little theory or guidance on how to choose *a priori* which recombination operator to use on a new problem. There is also very little guidance on how to choose how often to apply recombination (often referred to as the *recombination rate*). There have been three approaches to this problem, referred to as *static*, *predictive*, and *adaptive* approaches.

19.4.1 Static techniques

The simplest approach is to assume that one particular recombination operator should be applied at some static rate for all problems. The static rate is estimated from a set of empirical studies, over a wide variety of problems, population sizes, and mutation rates. Three studies are of note. De Jong (1975) studied the on-line and off-line performance of a GA on the De Jong test suite and recommended a recombination rate of 60% for one-point recombination (i.e. 60% of the population should undergo recombination). Grefenstette (1986) studied on-line performance of a GA on the De Jong test suite and recommended that one-point recombination be used at the higher rate of 95%. In the most recent study, Eshelman *et al* (1989) studied the mean number of evaluations required to find the global optimum on the De Jong test suite and recommended an intermediate rate of 70% for one-point recombination. Each of these settings for the recombination rate is associated with particular settings for mutation rates and population sizes, so the interested reader is encouraged to consult these references for more complete information.

19.4.2 Predictive techniques

In the static approach it is assumed that some fixed recombination rate (and recombination operator) is reasonable for a large number of problems. However, this will not be true in general. The predictive approaches are designed to predict the performance of recombination operators (i.e. to recognize when the recombination bias is correct or incorrect for the problem at hand).

Manderick *et al* (1991) computed fitness correlation coefficients for different recombination operators on various problems. Since they noticed a high correlation between operators with high correlation coefficients and good GA performance their approach was to choose the recombination operator with the highest correlation coefficient. The approach of Grefenstette (1995) was similar in spirit to that of Manderick *et al*. Grefenstette used a *virtual* GA to compute the past performance of an operator as an estimate of the future performance of an operator. By running the virtual GA with different recombination operators, Grefenstette estimated the performance of those operators in a real GA.

Altenberg (1994) and Radcliffe (1994) have proposed different predictive measures. Altenberg proposed using an alternative statistic referred to as the *transmission function in the fitness domain.* Radcliffe proposed using the *fitness variance of formae* (generalized schemata). Thus far all approaches have shown considerable promise.

19.4.3 Adaptive techniques

In both the static and predictive approaches the decision as to which recombination operator and the rate at which it should be applied is made prior to actually running the EA. However, since these approaches can make errors (i.e. choose nonoptimal recombination operators or rates), a natural solution is to make these choices adaptive. Adaptive approaches are designed to recognize when bias is correct or incorrect, and recover from incorrect biases when possible. For the sake of exposition adaptive approaches will be divided into *tag-based* and *rule-based*. As a general rule, tag-based approaches attach extra information to a chromosome, which is both evolved by the EA and used to control recombination. The rule-based approaches generally adapt recombination using control mechanisms and data structures that are external to the EA.

One of the earliest tag-based approaches was that of Rosenberg (1967). In this approach integers x_i ranging from zero to seven were attached to each locus. The recombination site was chosen from the probability distribution defined over these integers, $p_i = x_i / \sum x_i$, where p_i represented the probability of a cross at site i.

Schaffer and Morishima (1987) used a similar approach, by adjusting the points at which recombination was allowed to cut and splice material. They accomplished this by appending an additional L bits to L-bit individuals. These appended bits were used to determine *cut points* for each locus (a 'one' denoted a cut point while a 'zero' indicated the lack of a cut point). If two individuals had n distinct cut points, this was analogous to using a particular instantiation of n-point recombination.

Levenick (1995) also had a similar approach. Recombination was implemented by replicating two parents from one end to the other by iterating the following algorithm:

(i)   Copy one bit from parent1 to child1
(ii) Copy one bit from parent2 to child2
(iii) With some base probability P_b perform a recombination: swap the roles of the children (so subsequent bits come from the other parent).

Levenick inserted a metabit before each bit of the individual. If the metabit was 'one' in both parents recombination occurred with probability P_b, else recombination occurred with a reduced probability P_r. The effect was that the probability of recombination could be reduced from a maximum of P_b to a minimum of P_r. Levenick claimed that this method improved performance in those cases where the population did not converge too rapidly.

Arabas *et al* (1995) experimented with adaptive recombination in an evolution strategy. Each chromosome consisted of L real-valued parameters combined with an additional L control parameters. In a standard evolution strategy these extra control parameters are used to adapt mutation. In this particular study the control parameters were also used to adapt recombination, by concentrating offspring around particular parents. Empirical results on four classes of functions were encouraging.

Angeline (1996) evolved LISP expressions, and associated a recombination probability with each node in the LISP expressions. These probabilities evolved and controlled the application of recombination. Angeline investigated two different adaptive mechanisms based on this approach and reported that the adaptive mechanisms outperformed standard recombination on three test problems.

Spears (1995) used a simple approach in which one extra tag bit (Section 14.5) was appended to every individual. The tag bits were used to control the use of two-point and uniform recombination in the following manner:

if (parent1$[L + 1]$ = parent2$[L + 1]$ = 1)
then two-point-recombination(parent1, parent2)
else if (parent1$[L + 1]$ = parent2$[L + 1]$ = 0)
then uniform-recombination(parent1, parent2)
else if (rand$(0, 1)$ < 0.5)
    then two-point-recombination(parent1, parent2)
    else uniform-recombination(parent1, parent2)

Spears compared this adaptive approach on a number of different problems and population sizes, and found that the adaptive approach always had a performance intermediate between the best and worst of the two single recombination operators.

Rule-based approaches use auxiliary data structures and statistics to control recombination. The simplest of these approaches use hand-coded rules that associate various statistics with changes in the recombination rate. For example, Wilson (1986) examined the application of GAs to classifier systems and defined an entropy measure over the population. If the change in entropy was sufficiently positive or negative, the probability of recombination was decreased or increased respectively. The idea was to introduce more variation by increasing the recombination rate whenever the previous variation had been 'absorbed'. Booker (1987) considered the performance of GAs in function optimization and measured the percentage of the current population that produced offspring. Every percentage change in that measure was countered with an equal and opposite percentage change in the recombination rate.

Srinivas and Patnaik (1994) considered measures of fitness performance and used those measures to estimate the distribution of the population. They increased the probability of recombination (P_c) and the probability of mutation (P_m) when the population was stuck at local optima and decreased the probabilities when the population was scattered in the solution space. They also considered the need to preserve 'good' solutions of the population and attempted this by having lower values of P_c and P_m for high-fitness solutions and higher values of P_c and P_m for low-fitness solutions.

Hong *et al* (1995) had a number of different schemes for adapting the use of multiple recombination operators. The first scheme was defined by the following rule: if both parents were generated via the same recombination operator, apply that recombination operator, else randomly select (with a coin flip) which operator to use. This first scheme was very similar to that of Spears (1995) but does not use tag bits. Their second scheme was the opposite of the first: if both parents were generated via the same recombination operator, apply some other recombination operator, else randomly select (with a coin flip) which operator to use. Their third scheme used a measure called an *occupancy rate*, which was the number of individuals in a population that were generated by a particular recombination (divided by the population size). For k recombination operators, their third scheme tried to balance the occupancy rate of each recombination operator around $1/k$. In their experiments the second and third schemes outperformed the first (although this was not true when they tried uniform and two-point recombination).

Eshelman and Schaffer (1994) provided a switching mechanism to decide between two recombination operators that often perform well, HUX (a variant of uniform crossover where exactly half of the differing bits are swapped at random) and SHX (a version of one-point recombination in which the positional bias has been removed). Their GA uses restarts—when the population is (nearly) converged the converged population is partially or fully randomized and seeded with one copy of the best individual found so far (the *elite* individual). During any convergence period between restarts (including the period leading up to the first restart), either HUX or SHX is used but not both. HUX is always used during the first two convergences. Subsequently, three rules are used for switching recombination operators:

(i)   SHX is used for the next convergence if during the prior convergence no individual is found that is as good as the elite individual.

(ii) HUX is used for the next convergence if during the prior convergence no individual is found that is better than the elite individual, but at least one individual is found that is as good as the elite individual.

(iii) No change in the operator is made if during the prior convergence a new best individual is found (which will replace the old elite individual).

These methods had fairly simple rules and data structures. However, more complicated techniques have been attempted. Davis (1989) provided an elaborate bookkeeping method to reward recombination operators that produced good offspring or set the stage for this production. When a new individual was added to the population, a pointer was established to its parent or parents, and a pointer was established to the operator that created the new individual. If the new individual was better than the current best member of the population, the amount it was better was stored as its *local delta*. Local deltas were passed back to parents to produce *inherited deltas*. *Derived deltas* were the sums of local and inherited deltas. Finally, the *operator delta* was the sum of the derived deltas of the individuals it produced, divided by the number of individuals produced. These operator deltas were used to update the probability that the operator would be fired.

White and Oppacher (1994) used finite-state automata to identify groups of bits that should be kept together during recombination (an extension of uniform recombination). The basic idea was to learn from previous recombination operations in order to minimize the probability that highly fit schemata would be disrupted in subsequent recombination operations. The basic bitstring representation was augmented at each bit position with an automaton. Each state of the automaton mapped to a probability of recombination for that bitstring location—roughly, given N states, then the probability p_i associated with state i was i/N. Some of the heuristics used for updating the automaton state were:

(i)   if offspring fitness > fitness of the father(mother)
      then reward those bits that came from the father(mother)
(ii) if offspring fitness < fitness of the father(mother)
      then penalize those bits that came from the father(mother).

There were other rules to handle offspring of equal fitness. A reward implied that the automaton moved from state i to state $i + 1$, and a penalty implied that the automaton moved from state i to state $i - 1$.

Julstrom (1995) used an *operator tree* to fire recombination more often if it produced children of superior fitness. With each individual was an operator tree—a record of the operators that generated the individual and its ancestors. If a new individual had fitness higher than the current population median fitness, the individual's operator tree was scanned to compute the credit due to recombination (and mutation). A queue recorded the credit information for the most recent individuals. This information was used to calculate the probability of recombination (and mutation).

Finally, Lee and Takagi (1993) evolved fuzzy rules for GAs. The fuzzy rules had three input variables based on fitness measures:

(i) x = average fitness/best fitness
(ii) y = worst fitness/average fitness
(iii) z = change in fitness.

The rules had three possible outputs dealing with population size, recombination rate, and mutation rate. All of the variables could take on three values {small, medium, big}, with the semantics of those values determined by membership functions. The rules were evaluated by running the GA on the De Jong test suite and different rules were obtained for on-line and off-line performance. 51 rules were obtained in the fuzzy rulebase. Of these, 18 were associated with recombination. An example was: if (x is small) and (y is small) and (z is small) then the change in recombination rate is small.

In summary, the fixed-recombination-rate approaches are probably the least successful, but provide reasonable guesses for parameter settings. They also are reasonable settings for the initial stages of the predictive and adaptive approaches. The predictive approaches have had success and appear very promising. The adaptive approaches also have had some success. However, as Spears (1995) indicated, a common difficulty in the evaluation of the adaptive approaches has been the lack of adequate control studies. Thus, although the approaches may show signs of adaptation, it is not clear that adaptation is the cause of performance improvement.

19.5 Discussion

The successful application of recombination (or any other operator) involves a close link with the operator, the representation, and the objective function. This has been outlined by Peck and Dhawan (1995), who emphasized similarity—one needs to exploit similarities between previous high-performance samples, and these similar samples must have similar objective function values often enough for the algorithm to be effective. Goldberg (1989) and Falkenauer (1994) make a similar point when they refer to *meaningful building blocks*. This has led people to outline various issues that must be considered when designing a representation and appropriate recombination operators.

De Jong (1975) outlined several important issues with respect to representation. First, 'nearbyness' should be preserved, in which small changes in a parameter value should come about from small changes in the representation for that value. Thus, binary encodings of real-valued parameters are problematic, since Hamming cliffs separate parameters that are near in the real-valued space and standard recombination operators can produce offspring far removed (in the real-valued sense) from their parents. Second, it is generally better to have context-insensitive representations, in which the legal values for a parameter do not depend on the values of other parameters. Finally, it is generally better to have context-insensitive interpretations of the parameters, in which the interpretation of some parameter value does not depend on the values of the other parameters. These last two concerns often arise in permutation or ordering problems, in which the values of the leftmost parameters influence both the legal values and the interpretation of these values for the rightmost parameters. For example, the encoding of the TSP problem presented earlier is context sensitive, and standard recombination operators can produce invalid offspring when using the representation. An alternative representation could be one in which the first parameter specifies which of the N cities should be visited first. Having deleted that city from the list of cities, the second parameter always takes on a value in the range $1 \ldots N - 1$, specifying by position on the list which of the remaining cities is to be visited second, and so on. For example, suppose there are four cities {A, B, C, D}. The representation of the tour BCAD is (2 2 1 1) because city B is the second city in the list {A, B, C, D}, C is the second city in the list {A, C, D}, A is the first city in the list {C, D} and D is the first city in the list {D}. This representation is context insensitive and recombination of two tours always yields a valid tour. However, it has a context-sensitive interpretation, since gene values to the right of a recombination cut point specify different subtours in the parent and the offspring.

Radcliffe (1991) outlined three design principles for recombination. First, recombination operators should be respectful. Respect occurs if crossing two instances of any forma (a generalization of schema) must produce another instance of that forma. For example, if both parents have blue eyes then all their children must have blue eyes. This principle holds for any standard recombination on bitstrings. Second, recombination should *properly assort* formae. This occurs if, given instances of two compatible formae, it must be possible to cross them to produce an offspring which is an instance of both formae. For example, if one parent has blue eyes and the other has brown hair, it must be possible to recombine them to produce a child with blue eyes and brown hair. This principle is similar to what others called exploratory power— e.g. uniform recombination can reach all points in the subspace defined by the differing bits (in one application), while n-point recombination cannot. Thus n-point recombination does not properly assort, while uniform recombination does. Finally, recombination should *strictly transmit*. Strict transmission occurs if every allele in the child comes from one parent or another. For example, if one parent has blue eyes and the other has brown eyes, the child must have blue or brown eyes. All standard recombination operators for bitstrings strictly transmit genes.

All of this indicates that the creation and successful application of recombination operators is not 'cut and dried', nor a trivial pursuit. Considerable effort and thought is required. However, if one uses the guidelines suggested above as a first cut, success is more likely.

Acknowledgements

I thank Diana Gordon, Chad Peck, Mitch Potter, Ken De Jong, Peter Angeline, David Fogel, and the George Mason University GA Group for helpful comments on organizing this paper.

References

Altenberg L 1994 The schema theorem and Price's theorem *Proc. 3rd Foundations of Genetic Algorithms Workshop* ed M Vose and D Whitley (San Mateo, CA: Morgan Kaufmann) pp 23–49

Angeline P 1996 Two self-adaptive crossover operations for genetic programming *Adv. Genet. Programming* **2** 89–110

Arabas J, Mulawka J and Pokrasniewicz J 1995 A new class of the crossover operators for the numerical optimization *Proc. 6th Int. Conf. on Genetic Algorithms (Pittsburgh, PA, 1995)* ed L J Eshelman (San Mateo, CA: Morgan Kaufmann) 42–8

Beyer H-G 1995 Toward a theory of evolution strategies: on the benefits of sex—the $(\mu/\mu, \lambda)$ theory *Evolutionary Computation* **3** 81–112

Booker L B 1982 *Intelligent Behavior as an Adaptation to the Task Environment* PhD Dissertation, University of Michigan

——1987 Improving search in genetic algorithms *Genetic Algorithms and Simulated Annealing* ed L Davis (Los Altos, CA: Morgan Kaufmann) pp 61–73

——1992 Recombination distributions for genetic algorithms *Proc. 2nd Foundations of Genetic Algorithms Workshop* ed D Whitley (San Mateo, CA: Morgan Kaufmann) pp 29–44

Bui T and Moon B 1995 On multi-dimensional encoding/crossover *Proc. 6th Int. Conf. on Genetic Algorithms (Pittsburgh, PA, 1995)* ed L J Eshelman (San Mateo, CA: Morgan Kaufmann) pp 49–56

Davis L 1985 Applying adaptive algorithms in epistatic domains *Proc. Int. Joint Conf. on Artificial Intelligence*

——1989 Adapting operator probabilities in genetic algorithms *Proc. 3rd Int. Conf. on Genetic Algorithms (Fairfax, VA, 1989)* ed J Schaffer (San Mateo, CA: Morgan Kaufmann) pp 61–9

——1991 Hybridization and numerical representation *Handbook of Genetic Algorithms* ed L Davis (New York: Van Nostrand Reinhold) pp 61–71

De Jong K 1975 *Analysis of the Behavior of a Class of Genetic Adaptive Systems* PhD Dissertation, University of Michigan

De Jong K and Spears W 1990 An analysis of the interacting roles of population size and crossover in genetic algorithms *Proc. Int. Conf. on Parallel Problem Solving from Nature* ed H-P Schwefel and R Männer (Berlin: Springer) pp 38–47

——1992 A formal analysis of the role of multi-point crossover in genetic algorithms *Annals of Mathematics and Artificial Intelligence* (Switzerland: Baltzer) **5** **1** 1–26

De Jong K, Spears W and Gordon D 1993 Using genetic algorithms for concept learning *Machine Learning* **13** 161–88

Eshelman L and Schaffer D 1992 Real-coded genetic algorithms and interval-schemata *Proc. 2nd Foundations of Genetic Algorithms Workshop* ed D Whitley (San Mateo, CA: Morgan Kaufmann) pp 187–202

——1994 Productive recombination and propagating and preserving schemata *Proc. 3rd Foundations of Genetic Algorithms Workshop* ed M Vose and D Whitley (San Mateo, CA: Morgan Kaufmann) pp 299–313

Eshelman L, Caruana R and Schaffer D 1989 Biases in the crossover landscape *Proc. 3rd Int. Conf. on Genetic Algorithms (Fairfax, VA, 1989)* ed J Schaffer (San Mateo, CA: Morgan Kaufmann) pp 10–19

Falkenauer E 1994 A new representation and operators for genetic algorithms applied to grouping problems *Evolutionary Computation* (Cambridge, MA: MIT Press) **2 2** 123–44

Fogel L, Owens A and Walsh M 1966 *Artificial Intelligence through Simulated Evolution* (New York: Wiley)

Fraser A 1957 Simulation of genetic systems by automatic digital computers I Introduction *Aust. J. Biol. Sci.* **10** 484–91

Fujiki C and Dickinson J 1987 Using the genetic algorithm to generate lisp source code to solve the prisoner's dilemma *Proc. 2nd Int. Conf. on Genetic Algorithms* (Pittsburgh, PA, 1987) ed J J Grefenstette (Hillsdale, NJ: Erlbaum) pp 236–40

Goldberg D 1989 *Genetic Algorithms in Search Optimization and Machine Learning* (Reading, MA: Addison-Wesley)

Goldberg D and Lingle R 1985 Alleles loci and the traveling salesman problem *Proc. 1st Int. Conf. on Genetic Algorithms and their Applications (Pittsburgh, PA, 1985)* ed J J Grefenstette (Hillsdale, NJ: Erlbaum) pp 154–9

Grefenstette J 1986 Optimization of control parameters for genetic algorithms *IEEE Trans. Syst. Man Cybernet.* **SMC-16** 122–8

——1995 *Virtual Genetic Algorithms: First Results* Navy Center for Applied Research in AI Report AIC-95-013

Grefenstette J, Ramsey C and Schultz A 1990 Learning sequential decision rules using simulation models and competition *Machine Learning* **54** 355–81

Holland J 1975 *Adaptation in Natural and Artificial Systems* (Ann Arbor, MI: University of Michigan Press)

Hong I, Kahng A and Moon B 1995 Exploiting synergies of multiple crossovers: initial studies *Proc. IEEE Int. Conf. on Evolutionary Computation*

Julstrom B 1995 What have you done for me lately? adapting operator probabilities in a steady-state genetic algorithm *Proc. 6th Int. Conf. on Genetic Algorithms (Pittsburgh, PA, 1995)* ed L J Eshelman (San Mateo, CA: Morgan Kaufmann) pp 81–7

Kahng A and Moon B 1995 Towards more powerful recombinations *Proc. 6th Int. Conf. on Genetic Algorithms (Pittsburgh, PA, 1995)* ed L J Eshelman (San Mateo, CA: Morgan Kaufmann) pp 96–103

Koza J 1994 *Genetic Programming II: Automatic Discovery of Reusable Subprograms* (Cambridge, MA: MIT Press)

Lee M and Takagi H 1993 Dynamic control of genetic algorithms using fuzzy logic techniques *Proc. 5th Int. Conf. on Genetic Algorithms (Urbana-Champaign, IL, 1993)* ed S Forrest (San Mateo, CA: Morgan Kaufmann) pp 77–83

Levenick J 1995 Metabits: generic endogenous crossover control *Proc. 6th Int. Conf. on Genetic Algorithms (Pittsburgh, PA, 1995)* ed L J Eshelman (San Mateo, CA: Morgan Kaufmann) pp 88–95

Manderick B, de Weger M and Spiessens P 1991 The genetic algorithms and the structure of the fitness landscape *Proc. 4th Int. Conf. on Genetic Algorithms (San Diego, CA, 1991)* ed R K Belew and L B Booker (San Mateo, CA: Morgan Kaufmann) pp 143–50

Moon B and Bui T 1994 Analyzing hyperplane synthesis in genetic algorithms using clustered schemata *Parallel Problem Solving from Nature—III (Lecture Notes in Computer Science 806)* pp 108–18

Oliver I, Smith D and Holland J 1987 A study of permutation crossover operators on the traveling salesman problem *Proc. 2nd Int. Conf. on Genetic Algorithms (Pittsburgh, PA, 1987)* ed J J Grefenstette (Hillsdale, NJ: Erlbaum) pp 224–30

Peck C and Dhawan A 1995 Genetic algorithms as global random search methods: an alternative perspective *Evolutionary Computation* (Cambridge, MA: MIT Press) **3** **1** 39–80

Radcliffe N 1991 Forma analysis and random respectful recombination *Proc. 4th Int. Conf. on Genetic Algorithms (San Diego, CA, 1991)* ed R K Belew and L B Booker (San Mateo, CA: Morgan Kaufmann) pp 222–9

——1994 Fitness variance of formae and performance prediction *Proc. 3rd Foundations of Genetic Algorithms Workshop* ed M Vose and D Whitley (San Mateo, CA: Morgan Kaufmann) pp 51–72

Rosca J 1995 Genetic programming exploratory power and the discovery of functions *Proc. 4th Annu. Conf. on Evolutionary Programming (San Diego, CA, 1995)* ed J R McDonnell, R G Reynolds and D B Fogel (Cambridge, MA: MIT Press) pp 719–36

Rosenberg R 1967 *Simulation of Genetic Populations with Biochemical Properties* PhD Dissertation, University of Michigan

Schaffer J, Caruana R, Eshelman L and Das R 1989 A study of control parameters affecting on-line performance of genetic algorithms for function optimization *Proc. 3rd Int. Conf. on Genetic Algorithms (Fairfax, VA, 1989)* ed J D Schaffer (San Mateo, CA: Morgan Kaufmann) pp 51–60

Schaffer J and Eshelman K 1991 On crossover as an evolutionarily viable strategy *Proc. 4th Int. Conf. on Genetic Algorithms (San Diego, CA, 1991)* ed R K Belew and L B Booker (San Mateo, CA: Morgan Kaufmann) pp 61–8

Schaffer J and Morishima A 1987 An adaptive crossover distribution mechanism for genetic algorithms *Proc. 2nd Int. Conf. on Genetic Algorithms (Pittsburgh, PA, 1987)* ed J J Grefenstette (Hillsdale, NJ: Erlbaum) pp 36–40

Schwefel H-P 1995 *Evolution and Optimum Seeking* (New York: Wiley)

Smith S 1980 Flexible learning of problem solving heuristics through adaptive search *Proc. 8th Int. Conf. on Artificial Intelligence* pp 422–5

Spears W 1992 Crossover or Mutation? *Proc. 2nd Foundations of Genetic Algorithms Workshop* ed D Whitley (San Mateo, CA: Morgan Kaufmann) pp 221–37

——1995 Adapting crossover in evolutionary algorithms *Proc. 4th Ann. Conf. on Evolutionary Programming (San Diego, CA, 1995)* ed J R McDonnell, R G Reynolds and D B Fogel (Cambridge, MA: MIT Press) pp 367–84

Spears W and De Jong K 1991 On the virtues of parameterized uniform crossover *Proc. 4th Int. Conf. on Genetic Algorithms (San Diego, CA, 1991)* ed R K Belew and L B Booker (San Mateo, CA: Morgan Kaufmann) pp 230–6

Srinivas M and Patnaik L 1994 Adaptive probabilities of crossover and mutation in genetic algorithms *IEEE Trans. Syst. Man Cybernet.* **SMC-244** 656–67

Starkweather T, McDaniel S, Mathias K, Whitley D and Whitley C 1991 A comparison of genetic sequencing operators *Proc. 4th Int. Conf. on Genetic Algorithms (San Diego, CA, 1991)* ed R K Belew and L B Booker (San Mateo, CA: Morgan Kaufmann) pp 69–76

Syswerda G 1989 Uniform crossover in genetic algorithms *Proc. 3rd Int. Conf. on Genetic Algorithms (Fairfax, VA, 1989)* ed J D Schaffer (San Mateo, CA: Morgan Kaufmann) pp 2–9

——1992 Simulated crossover in genetic algorithms *Proc. 2nd Foundations of Genetic Algorithms Workshop* ed D Whitley (San Mateo, CA: Morgan Kaufmann) pp 239–55

White T and Oppacher F 1994 Adaptive crossover using automata *Proc. Parallel Problem Solving from Nature Conf.* ed Y Davidor, H-P Schwefel and R Männer (New York: Springer)

Wilson S 1986 *Classifier System Learning of a Boolean Function* Rowland Institute for Science Research Memo RIS-27r

Wright A 1991 Genetic algorithms for real parameter optimization *Proc. Foundations of Genetic Algorithms Workshop* ed G Rawlins (San Mateo, CA: Morgan Kaufmann) pp 205–18

20

Parameter control

*A E Eiben, Robert Hinterding and
Zbigniew Michalewicz*

The two major steps in applying any heuristic search algorithm to a particular problem are the specification of the representation and the evaluation (fitness) function. These two items form the bridge between the original problem context and the problem-solving framework. When defining an evolutionary algorithm one needs to choose its components, such as variation operators (mutation and recombination operators) that suit the representation, selection mechanisms for selecting parents and survivors, and an initial population. Each of these components may have parameters, such as the probability of mutation, the tournament size of selection, or the population size. The values of these parameters largely determine whether the algorithm will find a near-optimum solution, and whether it will find such a solution efficiently. Choosing the right parameter values, however, is a time-consuming task and considerable effort has gone into developing good heuristics for it.

Globally, we distinguish two major forms of setting parameter values: parameter *tuning* and parameter *control*. By parameter tuning we mean the commonly practised approach that amounts to finding good values for the parameters *before* the run of the algorithm and then running the algorithm using these values, which remain fixed during the run. Later we give arguments that any static set of parameters, having the values fixed during an EA run, seems to be inappropriate. Parameter control forms an alternative, as it amounts to starting a run with initial parameter values which are changed *during* the run.

In this chapter we provide a comprehensive discussion of parameter control and categorize different ways of doing it. The classification is based on two aspects: *how* the mechanism of change works, and *what* component of the EA is effected by the mechanism. Such a classification can be useful to the evolutionary computation community, since many researchers interpret terms such as 'adaptation' or 'self-adaptation' differently, which can be confusing. The framework we propose here is intended to eliminate ambiguities in the terminology. There are some other classification schemes, such as those of Angeline (1995), Hinterding *et al* (1997) or Smith and Fogarty (1997), that use other division criteria, resulting in different classification schemes.

The classification of Angeline (1995) is based on levels of adaptation and type of update rules. In particular, three levels of adaptation, population-level, individual-level, and component-level,† are considered, together with two types of update mechanism, absolute and empirical rules. Absolute rules are predetermined and specify how modifications should be made. On the other hand, empirical update rules modify parameter values by competition among them (self-adaptation). Angeline's framework considers an EA as a whole, without paying attention to its different components (e.g. mutation, recombination, selection, etc). The classification proposed by Hinterding *et al* (1997) extends that of Angeline (1995) by considering an additional level of adaptation (environment-level), and makes a more detailed division of types of update mechanism, dividing them into deterministic, adaptive, and self-adaptive categories. Here again, no attention is paid to what parts of an EA are adapted. The classification of Smith and Fogarty (1997) is probably the most comprehensive one. It is based on three division criteria: what is being adapted, the scope of the adaptation, and the basis for change. The last criterion is further divided into two categories: the evidence the change is based upon and the rule/algorithm that executes the change. Moreover, there are two types of rule/algorithm: uncoupled/absolute and tightly-coupled/empirical, the latter coinciding with self-adaptation.

The classification scheme discussed in this chapter is based on the type of update mechanisms and the EA component that is adapted, as basic division criteria. This classification addresses the key issues of parameter control without getting lost in details.

During the 1980s, a standard genetic algorithm (GA) based on bit representation, one-point crossover, bit-flip mutation and roulette wheel selection (with or without elitism) was widely applied. Algorithm design was thus limited to choosing the so-called control parameters, or strategy parameters‡, such as mutation rate, crossover rate and population size. Many researchers based their choices on tuning the control parameters 'by hand'; that is, experimenting with different values and selecting the ones that gave the best results. Later, they reported their results of applying a particular EA to a particular problem, stating:

> ...for these experiments, we have used the following parameters:
> population size of 100, probability of crossover equal to 0.85, etc

without much justification of the choice made.

Two main approaches were tried to improve GA design in the past. First, De Jong (1975) put a considerable effort into finding parameter values (for a

---

† Notice that we use the term 'component' differently from Angeline: he denotes subindividual structures with it, while we refer to parts of an EA, such as mutation, recombination, and selection.
‡ By 'control parameters' or 'strategy parameters' we mean the parameters of the EA, not those of the problem.

traditional GA), which were good for a number of numeric test problems. He determined experimentally recommended values for the probabilities of using single-point crossover and bit mutation. His conclusions were that the following parameters give reasonable performance for his test functions (for new problems these values may not be very good):

> population size of 50
> probability of crossover equal to 0.6
> probability of mutation equal to 0.001
> generation gap of 100%
> scaling window: $n = \infty$
> selection strategy: elitist.

Grefenstette (1986), on the other hand, used a GA as a meta-algorithm to optimize values for the same parameters for both on-line and off-line performance† of the algorithm. The best set of parameters to optimize the on-line (off-line) performance of the GA were (the values to optimize the off-line performance are given in parenthesis):

> population size of 30 (80)
> probability of crossover equal to 0.95 (0.45)
> probability of mutation equal to 0.01 (0.01)
> generation gap of 100% (90%)
> scaling window: $n = 1$ ($n = 1$)
> selection strategy: elitist (non-elitist).

Note that in both of these approaches, an attempt was made to find the optimal and *general* set of parameters; in this context, the word 'general' means that the recommended values can be applied to a wide range of optimization problems. Formerly, genetic algorithms were seen as robust problem solvers that exhibit approximately the same performance over a wide range of problems (Goldberg 1989, p 6). The contemporary view on EAs acknowledges that specific problems (problem types) require their specific EA setups for satisfactory performance. Thus, the scope of 'optimal' parameter settings is necessarily a narrow one. Any quest for generally (near-)optimal parameter settings is lost *a priori*. This stresses the need for efficient and powerful techniques that help in finding good parameter settings for a given problem; in other words, the need for good parameter tuning methods.

As an alternative to tuning parameters before running the algorithm, controlling them during a run was realized quite early (e.g. mutation step sizes in the evolution strategy community). Analysis of the simple corridor and sphere problems led to Rechenberg's 1/5 success rule (Rechenberg 1973), where feedback was used to control the mutation step size. Later, self-adaptation of mutation was used, where the mutation step size and the preferred direction

---

† These measures were defined originally by De Jong (1975); the intuition is that on-line performance is based on monitoring the best solution in each generation, while off-line performance takes all solutions in the population into account.

of mutation were controlled without any direct feedback. For certain types of problem, self-adaptive mutation was very successful and its use spread to other branches of evolutionary computation.

As mentioned earlier, parameter tuning by hand is a common practice in evolutionary computation. Typically one parameter is tuned at a time, which may cause some suboptimal choices, since parameters interact in a complex way. Simultaneous tuning of more parameters, however, leads to an enormous number of experiments. The technical drawbacks to parameter tuning based on experimentation can be summarized as follows:

- the process of parameter tuning costs a lot of time, even if parameters are optimized one by one, regardless of their interactions;
- for a given problem the selected parameter values are not necessarily optimal, even if the effort made in setting them was significant.

Other options for designing a good set of static parameters for an evolutionary method to solve a particular problem include *parameter setting by analogy* and the use of theoretical analysis. Parameter setting by analogy amounts to the use of parameter settings that have been proved successful for similar problems. However, it is not clear whether similarity between problems as perceived by the user implies that the optimal set of EA parameters is also similar. As for the theoretical approach, the complexities of evolutionary processes and characteristics of interesting problems allow theoretical analysis only after significant simplifications in either the algorithm or the problem model. Therefore, it is unclear what is the practical value of the current theoretical results on parameter settings. There are some theoretical investigations on the optimal population size (Goldberg 1989, Thierens 1996, Harik *et al* 1997, Goldberg *et al* 1992a) or optimal operator probabilities (Goldberg *et al* 1992b, Thierens and Goldberg 1991, Bäck 1993, Schaffer and Morishima 1987); however, these results were based on simple function optimization problems and their applicability for practical problems is limited.

A general drawback of the parameter tuning approach, regardless of how the parameters are tuned, is based on the observation that a run of an EA is an intrinsically dynamic, adaptive process. The use of rigid parameters that do not change their values is thus in contrast to the general evolutionary spirit. Additionally, it is intuitively obvious that different values of parameters might be optimal at different stages of the evolutionary process (Davis 1989, Syswerda 1991, Bäck 1992, 1993, 1994, Hesser and Männer 1991). For instance, large mutation steps can be good in the early generations for helping the exploration of the search space and small mutation steps might be needed in the late generations to help in fine-tuning the suboptimal chromosomes. This implies that the use of static parameters itself can lead to inferior algorithm performance. The straightforward way to treat this problem is by using parameters that may change over time, that is, by replacing a parameter p by a function $p(t)$, where t is the generation counter. However, as we indicated earlier, the problem of finding optimal *static* parameters for a particular problem can be quite difficult, and the optimal values may depend on many other factors (such as the applied recombination operator and the selection mechanism). Hence designing an optimal function $p(t)$ may be even more difficult. Another possible drawback to this approach is that the parameter value $p(t)$ changes are caused by a deterministic rule triggered by the progress of time t, without having any notion of the actual progress in solving the problem; that is, without taking into account the current state of the search. Yet researchers improved their evolutionary algorithms; that is, they improved the quality of results returned by their algorithms while working on particular problems, by using such simple deterministic rules. This can be explained simply by superiority of changing parameter values: suboptimal choice of $p(t)$ often leads to better results than a suboptimal choice of p.

To this end, recall that evolutionary algorithms implement the idea of evolution, and that evolution itself must have evolved to reach its current state of sophistication. It is thus natural to expect adaptation to be used not only for finding solutions to a problem, but also for tuning the algorithm to the particular problem. Technically speaking, this amounts to modifying the values of parameters during the run of the algorithm by taking the actual search process into account. Basically, there are two ways to do this. Either one can use some heuristic rule which takes feedback from the current state of the search and modifies the parameter values accordingly, or incorporate parameters into the chromosomes, thereby making them subject to evolution.

Let us assume we deal with a numerical optimization problem:

$$\text{optimize } f(\boldsymbol{x}) = f(x_1, \ldots, x_n)$$

subject to some inequality and equality constraints:

$$g_i(\boldsymbol{x}) \leq 0 \quad (i = 1, \ldots, q) \qquad \text{and} \qquad h_j(\boldsymbol{x}) = 0 \quad (j = q+1, \ldots, m)$$

and bounds $l_i \leq x_i \leq u_i$ for $1 \leq i \leq n$, defining the domain of each variable.

For such a numerical optimization problem we may consider an evolutionary algorithm based on floating-point representation. So, each individual $\boldsymbol{x}$ in the population is represented as a vector of floating-point numbers $\boldsymbol{x} = \langle x_1, \ldots, x_n \rangle$. Let us assume that we use Gaussian mutation together with arithmetical crossover to produce offspring for the next generation. A Gaussian mutation operator requires two parameters: the mean, which is supposed to be zero, and the standard deviation σ, which can be interpreted as the mutation step size. Mutations then are realized by replacing components of the vector $\boldsymbol{x}$ by

$$x_i' = x_i + N(0, \sigma)$$

where $N(0, \sigma)$ is a random Gaussian number with mean zero and standard deviation σ. The simplest (and the most popular) method to specify the mutation mechanism is to use the same σ for all vectors in the population, for all variables of each vector, and for the whole evolutionary process, for instance, $x_i' = x_i + N(0, 1)$. Intuitively, however, it might be beneficial to vary the mutation step size:† we shall discuss several possibilities in turn.

First, we can replace the static parameter σ by a dynamic parameter; that is, a function $\sigma(t)$. This function can be defined by some heuristic rule assigning different values depending on the number of generations. For example, the mutation step size may be defined as

$$\sigma(t) = 1 - 0.9\frac{t}{T}$$

where t is the current generation number; it varies from 0 to T, which is the maximum generation number. Here, the mutation step size $\sigma(t)$ (used for all for vectors in the population and for all variables of each vector) will decrease slowly from 1 at the beginning of the run ($t = 0$) to 0.1 as the number of generations t approaches T; such decreases may assist the fine-tuning capabilities of the algorithm. In this approach, the value of the given parameter changes according to a fully deterministic scheme; the user has thus full control of the parameter and its value at a given time t is completely determined and predictable.

Second, it is possible to incorporate feedback from the search process, still using the same σ for all for vectors in the population and for all variables of each vector. A well-known example of this type of parameter adaptation is Rechenberg's 1/5 success rule in (1+1)-evolution strategies (Rechenberg 1973). This rule states that the ratio of successful mutations‡ to all mutations should be 1/5, hence if the ratio is greater than 1/5 then the step size should be increased, and if the ratio is less than 1/5, the step size should be decreased:

> **if** $(t \bmod n = 0)$ **then**
> $$\sigma(t) := \begin{cases} \sigma(t-n)/c & \textbf{if } p_s > 1/5 \\ \sigma(t-n) \cdot c & \textbf{if } p_s < 1/5 \\ \sigma(t-n) & \textbf{if } p_s = 1/5 \end{cases}$$
> **else**
> $$\sigma(t) := \sigma(t-1);$$
> **fi**

where p_s is the relative frequency of successful mutations measured over some number of generations and $0.817 \le c \le 1$ (Bäck 1996). Using this mechanism, changes in the parameter values are now based on feedback from the search, and σ-adaptation happens every n generations. The influence of the user on the parameter values is much less here than in the deterministic scheme above.

---

† There are even formal arguments supporting this view in specific cases (e.g. Bäck 1992, 1993, 1994, Hesser and Männer 1991.

‡ A mutation is considered successful if it produces an offspring better than the parent.

Of course, the mechanism that embodies the link between the search process and parameter values is still a man-made heuristic rule telling how the changes should be made, but the values of $\sigma(t)$ are not predictable.

Third, it is possible to assign a 'personal' mutation step size to each individual: extend the representation to individuals of length $n + 1$ as $\langle x_1, \ldots, x_n, \sigma \rangle$, and apply some variation operators (e.g. Gaussian mutation and arithmetical crossover) to the x_i as well as to the σ value of an individual. In this way, not only the solution vector values (the x_i), but also the mutation step size of an individual undergoes evolution. A typical variation would be:

$$\sigma' = \sigma \, e^{N(0, \tau_0)} \qquad \text{and} \qquad x_i' = x_i + N(0, \sigma')$$

where τ_0 is a parameter of the method. This mechanism is commonly called self-adapting the mutation step sizes. Observe that within the self-adaptive scheme the heuristic character of the mechanism re-setting the parameter values is eliminated.†

Note that in the above scheme the scope of application of a certain value of σ was restricted to a single individual. However, it can be applied to all components of the individual: it is possible to change the granularity of such applications and use a separate mutation step size for each x_i. If an individual is represented as $\langle x_1, \ldots, x_n, \sigma_1, \ldots, \sigma_n \rangle$, then mutations can be realized by replacing the above vector according to a formula similar to that discussed above:

$$\sigma_i' = \sigma_i e^{N(0, \tau_0)} \qquad \text{and} \qquad x_i' = x_i + N(0, \sigma_i')$$

where τ_0 is a parameter of the method. However, as opposed to the previous case, each component x_i has its own mutation step size σ_i, which is being self-adapted. This mechanism implies a larger degree of freedom for adapting the search strategy to the topology of the fitness landscape.

So far we have described different ways to modify a parameter controlling mutation. Several other components of an EA have natural parameters, and these parameters are traditionally tuned in one or another way. Now we show that other components, such as the evaluation function (and consequently the fitness function) can also be parameterized and thus tuned. Although this is a less common option than tuning mutation, it may provide a powerful mechanism for increasing the performance of an evolutionary algorithm.

When dealing with constrained optimization problems, penalty functions are often used. The most popular one (Michalewicz and Schoenauer 1996) is the method of static penalties, which requires fixed user-supplied penalty parameters. The main reason for its popularity is that it is the simplest technique to implement: it requires only the straightforward modification of the evaluation

---

† It can be argued that the heuristic character of the mechanism re-setting the parameter values is not eliminated, but rather replaced by a meta-heuristic of evolution itself. However, the method is very robust with respect to the setting of τ_0 and a good rule is $\tau_0 = 1/\sqrt{n}$.

function *eval* as follows:

$$eval(\boldsymbol{x}) = f(\boldsymbol{x}) + W\ penalty(\boldsymbol{x})$$

where f is the objective function, and $penalty(\boldsymbol{x})$ is zero if no violation occurs, and is positive† otherwise. Usually, the *penalty* function is based on the distance of a solution from the feasible region, or on the effort to 'repair' the solution; that is, to force it into the feasible region. The first alternative is the most popular one; in many methods a set of functions f_j ($1 \le j \le m$) is used to construct the penalty, where the function f_j measures the violation of the jth constraint in the following way:

$$f_j(\boldsymbol{x}) = \begin{cases} \max\{0, g_j(\boldsymbol{x})\} & \text{if } 1 \le j \le q \\ |h_j(\boldsymbol{x})| & \text{if } q + 1 \le j \le m. \end{cases}$$

W is a user-defined weight, prescribing how severely constraint violations are weighted.‡ In the most traditional penalty approach the weight W does not change during the evolution process. We sketch three possible methods of changing the value of W.

First, we can replace the static parameter W by a dynamic parameter; that is, a function $W(t)$. Just as for the mutation parameter σ, we can develop a heuristic rule which modifies the weight W over time. For example, in the method proposed by Joines and Houck (1994), the individuals are evaluated (at the iteration t) by a formula, where

$$eval(\boldsymbol{x}) = f(\boldsymbol{x}) + (Ct)^{\alpha}\ penalty(\boldsymbol{x})$$

where C and α are constants. Clearly,

$$W(t) = (Ct)^{\alpha}$$

and the penalty pressure grows with the evolution time.

Second, let us consider another option, which utilizes feedback from the search process. One example of such an approach was developed by Bean and Hadj-Alouane (1992), where each individual is evaluated by the same formula as before, but $W(t)$ is updated in every generation t in the following way:

$$W(t + 1) = \begin{cases} (1/\beta_1)W(t) & \text{if } b^i \in \mathcal{F} \text{ for all } t - k + 1 \le i \le t \\ \beta_2 W(t) & \text{if } b^i \in \mathcal{S} - \mathcal{F} \text{ for all } t - k + 1 \le i \le t \\ W(t) & \text{otherwise.} \end{cases}$$

In this formula, $\mathcal{S}$ is the set of all search points (solutions), $\mathcal{F} \subseteq \mathcal{S}$ is a set of all *feasible* solutions, b^i denotes the best individual, in terms of function *eval*,

---

† For minimization problems.
‡ Of course, instead of W it is possible to consider a vector of weights $\boldsymbol{w} = (w_1, \ldots, w_m)$ which are applied directly to violation functions $f_j(\boldsymbol{x})$; in such a case $penalty(\boldsymbol{x}) = \sum_{j=1}^{m} w_j f(\boldsymbol{x})$. The discussion in the remaining part of this chapter can be easily extended to this case.

in generation i, $\beta_1, \beta_2 > 1$ and $\beta_1 \neq \beta_2$ (to avoid cycling). In other words, the method decreases the penalty component $W(t+1)$ for the generation $t+1$ if all best individuals in the last k generations were feasible (i.e. in $\mathcal{F}$), and increases penalties if all best individuals in the last k generations were infeasible. If there are some feasible and infeasible individuals as best individuals in the last k generations, $W(t+1)$ remains without change.

Third, we could allow self-adaptation of the weight parameter, similarly to the mutation step sizes in the previous chapter. For example, it is possible to extend the representation of individuals into $\langle x_1, \ldots, x_n, W \rangle$, where W is the weight. The weight component W undergoes the same changes as any other variable x_i (e.g. Gaussian mutation, arithmetical crossover). However, it is unclear how the evaluation function can benefit from such self-adaptation. Clearly, the smaller weight W, the better an (infeasible) individual is, so it is unfair to apply different weights to different individuals within the same generation. It might be that a new weight can be defined (e.g. arithmetical average of all weights present in the population) and used for evaluation purpose; however, to our best knowledge, no one has experimented with such self-adaptive weights.

To this end, it is important to note the crucial difference between self-adapting mutation step sizes and constraint weights. Even if the mutation step sizes are encoded in the chromosomes, the evaluation of a chromosome is *independent* from the actual values of σ. That is,

$$eval(\langle \boldsymbol{x}, \boldsymbol{\sigma} \rangle) = f(\boldsymbol{x})$$

for any chromosome $\langle \boldsymbol{x}, \boldsymbol{\sigma} \rangle$. In contrast, if constraint weights are encoded in the chromosomes, then we have

$$eval(\langle \boldsymbol{x}, W \rangle) = f_W(\boldsymbol{x})$$

for any chromosome $\langle \boldsymbol{x}, W \rangle$. This enables the evolution to 'cheat' in the sense of making improvements by modifying the value of W instead of optimizing f and satisfying the constraints!

We have illustrated how the mutation operator and the evaluation function can be controlled (adapted) during the evolutionary process. The latter case demonstrates that not only the traditionally adjusted components, such as mutation, crossover, and selection, can be controlled by parameters, but other components of an evolutionary algorithm as well. Obviously, there are many components and parameters that can be changed and tuned for optimal algorithm performance. In general, the three options we sketched for the mutation operator and the evaluation function are valid for any parameter of an evolutionary algorithm, whether it is population size, mutation step, penalty coefficient, selection pressure, and such like.

The mutation example also illustrates the phenomenon of the *scope of a parameter*. Namely, the mutation step size parameter can have different domains of influence, which we call scope. Using the $\langle x_1, \ldots, x_n, \sigma_1, \ldots, \sigma_n \rangle$ model, a particular mutation step size applies only to one variable of one individual. Thus, the parameter σ_i acts on a subindividual level. In the $\langle x_1, \ldots, x_n, \sigma \rangle$ representation the scope of σ is one individual, whereas the dynamic parameter $\sigma(t)$ was defined to affect all individuals and thus has the whole population as its scope.

In classifying parameter control techniques of an evolutionary algorithm, many aspects can be taken into account; for example:

(i)   *what* is changed?   (e.g. representation, evaluation function, operators, selection process, mutation rate, etc);

(ii) *how* is the change made?   (i.e. deterministic heuristic, feedback-based heuristic, or self-adaptive);

(iii) the *scope/level* of change (e.g. population-level, individual-level, etc);

(iv) the *evidence* upon which the change is carried out (e.g. monitoring performance of operators, diversity of the population, etc);

and used to classify the method accordingly. In the following we discuss these items in more detail.

To classify parameter control techniques from the perspective of 'what is changed?', it is necessary to agree on a list of *all* components of an evolutionary algorithm (which is a difficult task in itself). For example, one can assume the following components of an EA:

- representation of individuals,
- evaluation function,
- variation operators and their probabilities,
- selection operator (parent selection or mating selection),
- replacement operator (survival selection or environmental selection), and
- population (size, topology, etc).

Additionally, each component can be parameterized, and the number of parameters is not clearly defined. For example, an offspring produced by an arithmetical crossover of k parents $\boldsymbol{x}_1, \ldots, \boldsymbol{x}_k$ can be defined by the following formula

$$\boldsymbol{v} = a_1 \boldsymbol{x}_1 + \ldots + a_k \boldsymbol{x}_k$$

where $a_1, \ldots, a_k$, and k can be considered as parameters of this crossover. Parameters for population can include the number and sizes of subpopulations, migration rates, etc. (this is for a general case, when more then one population is involved). Despite these drawbacks, the 'what aspect' should be maintained as one of the main classification features, as it allows us to locate where a specific mechanism has its effect.

As mentioned earlier, each method for changing the value of a parameter (i.e. the 'how aspect') can be classified into one of three categories:

- *Deterministic parameter control.*
  This takes place when the value of a strategy parameter is altered by some deterministic rule; this rule modifies the strategy parameter deterministically without using any feedback from the search. Usually, a time-varying schedule is used; that is, the rule will be used when a set number of generations have elapsed since the last time the rule was activated.

- *Adaptive parameter control.*
  This takes place when there is some form of feedback from the search that is used to determine the direction and/or magnitude of the change to the strategy parameter. The assignment of the value of the strategy parameter may involve credit assignment, and the action of the EA may determine whether or not the new value persists or propagates throughout the population.

- *Self-adaptive parameter control.*
  The idea of the evolution of evolution can be used to implement the self-adaptation of parameters. Here the parameters to be adapted are encoded onto the chromosome(s) of the individual and undergo mutation and recombination. The 'better' values of these encoded individuals lead to 'better' individuals, which in turn are more likely to survive and produce offspring and hence propagate these 'better' parameter values.

Some authors have introduced different terminologies. Angeline (1995) distinguishes absolute and empirical rules corresponding to uncoupled and tightly coupled mechanisms of Spears (1995). Let us note that the uncoupled/absolute category encompasses deterministic and adaptive control, whereas the tightly coupled/empirical category corresponds to self-adaptation. We feel that the distinction between deterministic and adaptive parameter control is essential, as the former does not use any feedback from the search process.

As discussed earlier, any change (within any component of an EA) may affect a gene, whole chromosomes (individuals), the whole population, or even the evaluation function. This is the aspect of the scope or level of adaptation (Angeline 1995, Hinterding *et al* 1997, Smith and Fogarty 1997). Note, however, that the scope/level usually depends on the component of EA where the change takes place; for example, a change of the mutation step size may affect a gene, a chromosome, or the whole population, depending on the particular implementation (i.e. on the scheme used). On the other hand, a change in the penalty coefficients always affects the whole population. So, the scope/level feature is a secondary one, usually depending on the given component and its actual implementation.

The issue of the scope of the parameter might be more complicated than indicated earlier in this chapter, however. First of all, the scope depends on the interpretation mechanism of the given parameters. For example, an individual in evolution strategies might be represented as

$$\langle x_1, \ldots, x_n, \sigma_1, \ldots, \sigma_n, \alpha_1, \ldots, \alpha_{n(n-1)/2} \rangle$$

where the vector α denotes the covariances between the variables $\sigma_1, \ldots, \sigma_n$. In this case the scope of the strategy parameters in α is the whole individual, although the notation might suggest that they act on a subindividual level.

The next example illustrates that the same parameter (encoded in the chromosomes) can be interpreted in different ways, leading to different algorithm variants with different scopes of this parameter. Spears (1995) experimented with individuals containing an extra bit to determine whether one-point crossover or uniform crossover is to be used (bit 1/0 standing for one-point/uniform crossover, respectively). Two interpretations were considered. The first interpretation was based on a pairwise operator choice: if both parental bits are the same the corresponding operator is used, otherwise a random choice is made. Thus, this parameter in this interpretation acts at an individual level. The second interpretation was based on the bit distribution over the whole population: if, for example 73% of the population had bit 1, then the probability of one-point crossover was 0.73. Thus this parameter in this interpretation acts on population-level. Note, that these two interpretations can be easily combined. For instance, similar to the first interpretation, if both parental bits are the same, the corresponding operator is used. However, if they differ, the operator is selected according to the bit distribution, just as in the second interpretation. The scope/level of this parameter in this interpretation is neither individual, nor population, but rather both. This example shows that the notion of scope can be ill-defined and arbitrarily complex. This example and the arguments presented earlier about the scope of the parameter motivate our decision to exclude it as a major classification criterion.

Another possible criterion for classification is the evidence used for determining the change of parameter value (Smith and Fogarty 1997). Most commonly, the progress of the search is monitored (e.g. the performance of operators). It is also possible to look at other measures, such as the diversity of the population. The information gathered by such a monitoring process is used as feedback for adjusting the parameters. Although this is a meaningful distinction, it appears only in adaptive parameter control. A similar distinction can be made in deterministic control, which might be based on any counter not related to search progress. One option is the number of fitness evaluations (as the description of deterministic control above indicates). There are, however, other possibilities, for instance, changing the probability of mutation on the basis of the number of executed mutations. We feel, however, that these distinctions are of too low a level (in comparison with other criteria) and for that reason we have not included this as a major classification criterion.

So the main criteria for classifying methods which change the values of the strategy parameters of an algorithm during its execution are:

(i)   *what* is changed?

(ii) *how* is the change made?

The proposed classification is thus two-dimensional: the type of control and the component of the evolutionary algorithm which incorporates the parameter are taken into account. The *type* and *component* entries are orthogonal and encompass typical forms of parameter control within EAs. The *type* of parameter change consists of three categories: deterministic, adaptive, and self-adaptive mechanisms. The *component* of parameter change consists of six categories: representation, evaluation function, variation operators (mutation and recombination), selection, replacement, and population.

Summary

The effectiveness of an evolutionary algorithm depends on many of its components, such as representation and operators, and on the interactions among them. The variety of parameters included in these components, the many possible choices (e.g. to change or not to change?), and the complexity of the interactions between various components and parameters make the selection of a 'perfect' EA for a given problem very difficult, if not impossible.

So, how can we find the 'best' EA for a given problem? As discussed earlier in this chapter, we can perform a certain amount of parameter tuning, trying to find good values for all parameters before the run of the algorithm. However, even if we assume for a moment that there is a perfect configuration, finding it is an almost hopeless task. Figure 20.1 illustrates this point: the search space S_{EA} of all possible evolutionary algorithms is huge, much larger than the search space S_P of the given problem P, so our chances of *guessing* the right configuration (if one exists!) for an EA are rather slim (e.g. much smaller than the chances of guessing the optimum permutation of cities for a large instance of the traveling salesman problem). Even if we restrict our attention to a relatively narrow subclass, say S_{GA} of classical GAs, the number of possibilities is still prohibitive.† Note, that within this (relatively small) class there are many possible algorithms with different population sizes, different frequencies of the two basic operators (whether static or dynamic), and so on. Besides, guessing the right values of parameters might be of limited value anyway: in this chapter we have argued that any set of static parameters seems to be inappropriate, as any run of an EA is an intrinsically dynamic, adaptive process. So the use of rigid parameters that do not change their values may not be optimal, since different values of parameters may work better/worse at different stages of the evolutionary process.

On the other hand, adaptation provides the opportunity to customize the evolutionary algorithm to the problem and to modify the configuration and the

---

† A subspace of *classical* genetic algorithms, $S_{GA} \subset S_{EA}$, consists of evolutionary algorithms where individuals are represented by binary coded fixed-length strings, which has two operators: one-point crossover and a bit-flip mutation, and it uses a proportional selection.

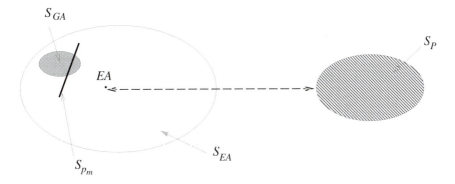

Figure 20.1. An evolutionary algorithm EA for problem P as a single point in the search space S_{EA} of all possible evolutionary algorithms. EA searches (broken line) the solution space S_P of the problem P. S_{GA} represents a subspace of classical GAs, whereas S_{p_m} represents a subspace consisting of evolutionary algorithms that are identical except for their mutation rate p_m.

strategy parameters used while the problem solution is sought. This possibility enables us not only to incorporate domain information and multiple reproduction operators into the EA more easily, but, as indicated earlier, allows the algorithm itself to select those values and operators that provide better results. Of course, these values can be modified during the run of the EA to suit the situation during that part of the run. In other words, if we allow some degree of adaptation within an EA, we can talk about two different searches which take place simultaneously: while the problem P is being solved (i.e. the search space S_P is being searched), a part of S_{EA} is searched as well for the best evolutionary algorithm EA for some stage of the search of S_P. However, in all experiments reported by various researchers only a tiny part of the search space S_{EA} was considered. For example, by adapting the mutation rate p_m we consider only a subspace S_{p_m} (see figure 20.1), which consists of all evolutionary algorithms with all parameters fixed except the mutation rate. Similarly, early experiments of Grefenstette (1986) were restricted to the subspace S_{GA} only.

An important objective of this paper is to draw attention to the potentials of EAs adjusting their own parameters on-line. Given the present state of the art in evolutionary computation, what could be said about the feasibility and the limitations of this approach?

One of the main obstacles of optimizing parameter settings of EAs is formed by the epistasic interactions between these parameters. The mutual influence of different parameters on each other and the combined influence of parameters on EA behavior is very complex. A pessimistic conclusion would be that such an approach is not appropriate, since the ability of EAs to cope with epistasis is limited. On the other hand, parameter optimization falls in the category of ill-defined, not well-structured (at least not well-understood) problems preventing an analytical approach—a problem class for which EAs usually provide a reasonable alternative to other methods. Roughly speaking, we might not have a better way to do it than letting the EA figure it out. To this end, note that the self-adaptive approach represents the highest level of reliance on the EA itself in setting the parameters. With a high confidence in the capability of EAs to solve the problem of parameter setting this is the best option. A more skeptical approach would provide some assistance in the form of heuristics on how to adjust parameters, amounting to adaptive parameter control. At this moment there are not enough experimental or theoretical results available to make any reasonable conclusions on the (dis)advantages of different options.

A theoretical boundary on self-adjusting algorithms in general is formed by the no-free-lunch theorem. However, while the theorem certainly applies to a self-adjusting EA, it represents a statement about the performance of the self-adjusting features in optimizing parameters compared to other algorithms for the same task. Therefore, the theorem is not relevant in the practical sense, because these other algorithms hardly exist in practice. Furthermore, the comparison should be drawn between the self-adjusting features and the human 'oracles' setting the parameters, this latter being the common practice.

It could be argued that relying on human intelligence and expertise is the best way of drawing an EA design, including the parameter settings. After all, the 'intelligence' of an EA would always be limited by the small fraction of the predefined problem space it encounters during the search, while human designers (may) have global insight of the problem to be solved. This, however, does not imply that the human insight leads to better parameter settings. Furthermore, human expertise is costly and might not be easily available for the given problem at hand, so relying on computer power is often the most practicable option. The domain of applicability of the evolutionary problem solving technology as a whole could be significantly extended by EAs that are able to configurate themselves, at least partially.

At this stage of research it is unclear just 'how much parameter control' might be useful. Is it feasible to consider the whole search space S_{EA} of evolutionary algorithms and allow the algorithm to select (and change) the representation of individuals together with operators? At the same time should the algorithm control probabilities of the operators used together with population size and selection method? It seems that more research on the combination of the types and levels of parameter control needs to be done. Clearly, this could lead to significant improvements to finding good solutions and to the speed of finding them.

Another aspect of the same issue is 'how much parameter control is worthwhile?'. In other words, what computational costs are acceptable? Some researchers have offered that adaptive control substantially complicates the task of EA and that the rewards in solution quality are not significant to justify the cost. Clearly, there is some learning cost involved in adaptive and self-adaptive control mechanisms. Either some statistics are collected during the run, or additional operations are performed on extended individuals. Comparing the efficiency of algorithms with and without (self-)adaptive mechanisms might be misleading, since it disregards the time needed for the tuning process. A fairer comparison could be based on a model that includes the time needed to set up (to tune) and to run the algorithm. We are not aware of any such comparisons at the moment.

On-line parameter control mechanisms may have a particular significance in nonstationary environments. In such environments often it is necessary to modify the current solution due to various changes in the environment (e.g. machine breakdowns, sickness of employees, etc). The capabilities of evolutionary algorithm to consider such changes and to track the optimum efficiently have been studied by several researchers. A few mechanisms were considered, including (self-)adaptation of various parameters of the algorithm, while other mechanisms were based on maintenance of genetic diversity and on redundancy of genetic material. These mechanisms often involved their own adaptive schemes (e.g. adaptive dominance function).

It seems that there are several exciting research issues connected with parameter control of EAs. These include:

- Developing models for comparison of algorithms with and without (self-) adaptive mechanisms. These models should include stationary and dynamic environments.

- Understanding the merit of parameter changes and interactions between them using simple deterministic controls. For example, one may consider an EA with a constant population size versus an EA where population size decreases, or increases, at a predefined rate such that the total numbers of function evaluations in both algorithms remain the same (it is relatively easy to find heuristic justifications for both scenarios).

- Justifying popular heuristics for adaptive control; for instance, why and how to modify mutation rates when the allele distribution of the population changes?

- Trying to find the general conditions under which adaptive control works. For self-adative mutation step sizes there are some universal guidelines (e.g. surplus of offspring, extinctive selection), but so far we do not know of any results regarding adaptation.

- Understanding the interactions among adaptively controlled parameters. Usually feedback from the search triggers changes in one of the parameters of the algorithm. However, the same trigger can be used to change the values of other parameters. The parameters can also directly influence each other.

- Investigating the merits and drawbacks of self-adaptation of several (possibly all) parameters of an EA.

- Developing a formal mathematical basis for the proposed taxonomy for parameter control in evolutionary algorithms in terms of functionals which transform the operators and variables they require.

In the next few years we expect new results in these areas.

References

Angeline P J 1995 Adaptive and self-adaptive evolutionary computation *Computational Intelligence: A Dynamic System Perspective* ed M Palaniswami, Y Attikiouzel, R J Marks, D Fogel and T Fukuda (Piscataway, NJ: IEEE)

Bäck T 1992 The interaction of mutation rate, selection, and self-adaptation within a genetic algorithm *Proc. 2nd Conf. on Parallel Problem Solving from Nature* ed R Männer and B Manderick (Amsterdam: North-Holland) pp 85–94

——1993 Optimal mutation rates in genetic search *Proc. 5th Int. Conf. on Genetic Algorithms (Urbana-Champaign, IL, July 1993)* ed S Forrest (San Mateo, CA: Morgan Kaufmann) pp 2–8

——1994 Self-adaption in genetic algorithms *Toward a Practice of Autonomous Systems: Proc. 1st Eur. Conf. on Artificial Life* ed F J Varela and P Bourgine (Cambridge, MA: MIT Press) pp 263–71

——1996 *Evolutionary Algorithms in Theory and Practice* (New York: Oxford University Press)

Bean J C and Hadj-Alouane A B 1992 A dual genetic algorithm for bounded integer programs *Technical Report 92-53* Department of Industrial and Operations Engineering, University of Michigan

Beasley D, Bull D R and Martin R R 1993 An overview of genetic algorithms: Part 2, Research topics *University Comput.* **15** pp 170–81

Davis L 1989 Adapting operator probabilities in genetic algorithms *Proc. 3rd Int. Conf. on Genetic Algorithms (Fairfax, VA, June 1989)* ed J D Schaffer (San Mateo, CA: Morgan Kaufmann) pp 61–9

De Jong K 1975 The analysis of the behavior of a class of genetic adaptive systems *PhD Dissertation* Department of Computer Science, University of Michigan

Goldberg D E 1989 *Genetic Algorithms in Search, Optimization and Machine Learning* (Reading, MA: Addison-Wesley)

Goldberg D E, Deb K, and Clark J H 1992a Genetic algorithms, noise, and the sizing of populations *Complex Systems* **6** 333–62

Goldberg D E, Deb K, and Thierens D 1992b Toward a better understanding of mixing in genetic algorithms *Proc. 4th Int. Conf. on Genetic Algorithms (San Diego, CA, July 1991)* ed R Belew and L Booker (San Mateo, CA: Morgan Kaufmann) pp 190–5

Grefenstette J J 1986 Optimization of control parameters for genetic algorithms *IEEE Trans. Systems, Man, Cybern.* **16** 122–8

Harik G, Cantu-Paz E, Goldberg D E, and Miller B L 1996 The gambler's ruin problem, genetic algorithms, and the sizing of populations *Proc. 4th IEEE Conf. on Evolutionary Computation* (Piscataway, NJ: IEEE) pp 7–12

Hesser J and Männer R 1991 Towards an optimal mutation probability for genetic algorithms *Proc. 1st Conf. on Parallel Problem Solving from Nature (Lecture Notes in Computer Science 496)* ed H-P Schwefel and R Männer (Berlin: Springer) pp 23–32

Hinterding R, Michalewicz Z, and Eiben A E 1997 Adaptation in evolutionary computation: a survey *Proc. 4th IEEE Conf. on Evolutionary Computation* (Piscataway, NJ: IEEE) pp 65–9

Joines J A and Houck C R 1994 On the use of non-stationary penalty functions to solve nonlinear constrained optimization problems with GAs *Proc. 1st IEEE Conf. on Evolutionary Computation* (Piscataway, NJ: IEEE) pp 579–84

Michalewicz Z and Schoenauer M 1996 Evolutionary algorithms for constrained parameter optimization problems *Evolutionary Comput.* **4** 1–32

Rechenberg R 1973 *Evolutionsstrategie: Optimierung technischer Syseme nach Prinzipien der biologischen Evolution* (Stuttgart: Frommann-Holzboog)

Schaffer J D and Morishima A 1987 An adaptive crossover distribution mechanism for genetic algorithms *Proc. 2nd Int. Conf. on Genetic Algorithms (Cambridge, MA, 1987)* ed J J Grefenstette (Hillsdale, NJ: Erlbaum) pp 36–40

Smith J E and Fogarty T C 1997 Operator and parameter adaptation in genetic algorithms *Soft Comput.* **1** 81–7

Spears W M 1995 Adapting crossover in evolutionary algorithms. *Proc. 4th Annual Conf. on Evolutionary Programming* ed J R McDonnell, R G Reynolds and D B Fogel (Cambridge, MA: MIT Press) pp 367–84

Syswerda G 1991 Schedule optimization using genetic algorithms *Handbook of Genetic Algorithms* ed L Davis (New York: Van Nostrand Reinhold) pp 332–49

Thierens D 1996 Dimensional analysis of allele-wise mixing revisited *Proc. 4th Conf. on Parallel Problem Solving from Nature (Lecture Notes in Computer Science 1141)* ed H-M Voigt, W Ebeling, I Rechenberg and H-P Schwefel (Berlin: Springer) pp 255–65

Thierens D and Goldberg D E 1991 Mixing in genetic algorithms *Proc. 4th Int. Conf. on Genetic Algorithms (San Diego, CA, July 1991)* ed R Belew and L Booker (San Mateo, CA: Morgan Kaufmann) pp 31–7

21

Self-adaptation

Thomas Bäck

21.1 Introduction

The *self-adaptation* of strategy parameters provides one of the key features of
the success of evolution strategies and evolutionary programming, because both
evolutionary algorithms use evolutionary principles to search in the space of
object variables and strategy parameters simultaneously.

The term *strategy parameters* refers to parameters that control the
evolutionary search process, such as mutation rates, mutation variances, and
recombination probabilities, and the idea of self-adaptation consists in evolving
these parameters in analogy to the object variables themselves. Typically,
strategy parameters are self-adapted on the level of individuals, by incorporating
them into the representation of individuals in addition to the set of object
variables; that is, the individual space I is given by

$$I = A_x \times A_s \tag{21.1}$$

where A_x denotes the set of object variables (i.e. of representations of solutions)
and A_s denotes the set of strategy parameters.

For an individual $a = (x, s)$ consisting of an object variable vector x and a
strategy parameter set s, the self-adaptation mechanism is typically implemented
by first (recombining and) mutating (according to some probability density
function) the strategy parameter vector s, yielding s', and then using the updated
strategy parameters s' to (recombine and) mutate the object variable vector x,
yielding x'.

Consequently, rather than using some deterministic control rule for the
modification of strategy parameters, they are themselves subject to evolutionary
operators and probabilistic changes. Selection is still performed on the basis
of the objective function value $f(x)$ only; that is, strategy parameters are
selected for survival by means of the *indirect link* between strategy parameters
and the objective function value. Since the mechanism works on the basis
of rewarding improvements in objective function value, strategy parameters
are continuously adapted such that convergence velocity is emphasized by the

evolutionary algorithm, but the speed of the adaptation on the level of strategy parameters is under the control of the user by means of so-called *learning rates*.

It should be noted that the self-adaptation principle is fundamentally different from other parameter control mechanisms for evolutionary algorithms such as *dynamic parameter control* or *adaptive parameter control*—a classification that was recently proposed by Eiben and Michalewicz (1996). Under dynamic parameter control, the parameter settings obtain different values according to a deterministic schedule prescribed by the user. An overview of dynamic schedules can be found in Chapter 18. Adaptive parameter control mechanisms obtain new values by a feedback mechanism that monitors evolution and explicitly rewards or punishes operators according to their impact on the objective function value. Examples of this mechanism are the method of Davis (1989) to adapt operator probabilities in genetic algorithms based on their observed success or failure to yield a fitness improvement and the approaches of Arabas *et al* (1994) and Schlierkamp-Voosen and Mühlenbein (1996) to adapt population sizes either by assigning lifetimes to individuals based on their fitness or by having a competition between subpopulations based on the fitness of the best population members. In contrast to these approaches, *self-adaptive parameter control* works by encoding parameters in the individuals and evolving the parameters themselves.

The following sections give an overview of some of the approaches for self-adaptation of strategy parameters described in the literature.

21.2 Mutation operators

Most of the research and successful applications of self-adaptation principles in evolutionary algorithms deal with parameters related to the mutation operator. The technique of self-adaptation is most widely utilized for the variances and covariances of a generalized n-dimensional normal distribution, as introduced by Schwefel (1977) in the context of evolution strategies and Fogel (1992) for the parameter optimization variants of evolutionary programming. The case of continuous object variables $x_i \in \mathbb{R}$ motivated a number of successful recent attempts to transfer the method to other search spaces such as binary vectors, discrete spaces in general, and even finite-state machines. In the following subsections, the corresponding self-adaptation principles are described in some detail.

21.2.1 Continuous search spaces

In the most general case, an individual $a = (x, \sigma, \alpha)$ of a (μ, λ) evolution strategy consists of up to three components $x \in \mathbb{R}^n$, $\sigma \in \mathbb{R}^{n_\sigma}$, and $\alpha \in [-\pi, \pi]^{n_\alpha}$, where $n_\sigma \in \{1, \ldots, n\}$ and $n_\alpha \in \{0, (2n - n_\sigma)(n_\sigma - 1)/2\}$. The mutation operator works by adding a realization of a normally distributed

n-dimensional random variable $X \sim N(0, C)$ with expectation vector 0, covariance matrix

$$C = (c_{ij}) = \begin{cases} \text{cov}(X_i, X_j) & i \neq j \\ \text{var}(X_i) & i = j \end{cases} \tag{21.2}$$

and probability density function

$$f_X(x_1, \ldots, x_n) = \frac{\exp\left(-\frac{1}{2}x^T C^{-1} x\right)}{((2\pi)^n \det(C))^{1/2}} \tag{21.3}$$

where the covariance matrix is described by the mutated strategy parameters σ' and α' of the individual. Depending on the number of strategy parameters incorporated into the representation of an individual, the following main variants of self-adaptation can be distinguished.

(i) $n_\sigma = 1$, $n_\alpha = 0$, $X \sim \sigma' N(0, I)$. The standard deviation for all object variables is identical (σ'), and all object variables are mutated by adding normally distributed random numbers with

$$\sigma' = \sigma \exp(\tau_0 N(0, 1)) \tag{21.4}$$
$$x_i' = x_i + \sigma' N_i(0, 1) \tag{21.5}$$

where $\tau_0 \propto n^{-1/2}$ and $N_i(0, 1)$ denotes a realization of a one-dimensional normally distributed random variable with expectation zero and standard deviation one that is sampled anew for each index i. The lines of equal probability density of the normal distribution are hyperspheres in this case, as shown graphically for $n = 2$ in the left-hand part of figure 21.1.

(ii) $n_\sigma = n$, $n_\alpha = 0$, $X \sim N(0, \sigma' I)$. All object variables have their own, individual standard deviations σ_i, which determine the corresponding modifications according to

$$\sigma_i' = \sigma_i \exp(\tau' N(0, 1) + \tau N_i(0, 1)) \tag{21.6}$$
$$x_i' = x_i + \sigma_i' N_i(0, 1) \tag{21.7}$$

where $\tau' \propto (2n)^{-1/2}$ and $\tau \propto (2n^{1/2})^{-1/2}$. The lines of equal probability density of the normal distribution are hyperellipsoids, as shown in the middle part of figure 21.1 for $n = 2$.

(iii) $n_\sigma = n$, $n_\alpha = n(n-1)/2$, $X \sim N(0, C)$. The vectors σ and α represent the complete covariance matrix of the n-dimensional normal distribution, where the covariances c_{ij} ($i \in \{1, \ldots, n-1\}$, $j \in \{i+1, \ldots, n\}$) are represented by a vector of rotation angles α_k ($k = \frac{1}{2}(2n-i)(i+1) - 2n + j$) describing the coordinate rotations necessary to transform an uncorrelated mutation vector into a correlated one. Rotation angles and covariances are related to each other according to

$$\tan(2\alpha_k) = \frac{2c_{ij}}{\sigma_i^2 - \sigma_j^2}. \tag{21.8}$$

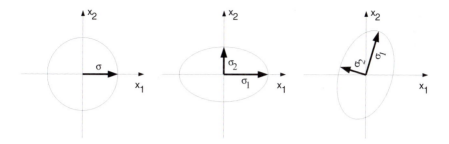

Figure 21.1. A sketch of the lines of equal probability density of the $n = 2$-dimensional normal distribution in the case of simple mutations with $n_\sigma = 1$ (left), $n_\sigma = 2$ (middle), and correlated mutations with $n_\sigma = 2$, $n_\alpha = 1$ (right).

By using the rotation angles to represent the covariances, the mutation operator is guaranteed to generate exactly the feasible (positive definite) covariance matrices and to allow for the creation of any possible covariance matrix (for details see the article by Rudolph (1992)). The mutation is performed according to

$$\sigma_i' = \sigma_i \exp(\tau' N(0, 1) + \tau N_i(0, 1)) \tag{21.9}$$
$$\alpha_j' = \alpha_j + \beta N_j(0, 1) \tag{21.10}$$
$$\boldsymbol{x}' = \boldsymbol{x} + \mathbf{N}(\mathbf{0}, \mathbf{C}(\sigma', \alpha')) \tag{21.11}$$

where $\mathbf{N}(\mathbf{0}, \mathbf{C}(\sigma', \alpha'))$ denotes the correlated mutation vector and $\beta \approx 0.0873$ ($5°$). As shown in the right-hand part of figure 21.1 for $n = 2$, the mutation hyperellipsoids are now arbitrarily rotatable, and α_k (with $k = \frac{1}{2}(2n - i)(i + 1) - 2n + j$) characterizes the rotation angle with respect to the coordinate axes i and j.

(iv) $1 < n_\sigma < n$. The general case of having neither just one nor the full number of different degrees of freedom available is also permitted, and implemented by the agreement to use σ_{n_σ} for mutating all x_i where $n_\sigma \le i \le n$.

The settings for the *learning rates* τ, τ', and τ_0 are recommended by Schwefel as reasonable heuristic settings (see Schwefel 1977, pp 167–8), but one should have in mind that, depending on the particular topological characteristics of the objective function, the optimal setting of these parameters might differ from the values proposed. For $n_\sigma = 1$, however, Beyer (1995b) has recently theoretically shown that, for the sphere model

$$f(\boldsymbol{x}) = \sum_{i=1}^{n} (x_i - x_i^*)^2 \tag{21.12}$$

the setting $\tau_0 \propto n^{-1/2}$ is the optimal choice, maximizing the convergence velocity of the evolution strategy. Moreover, for a $(1, \lambda)$ evolution strategy

Beyer derived the result that $\tau_0 \approx c_{1,\lambda}/n^{1/2}$ (for $\lambda \geq 10$), where $c_{1,\lambda}$ denotes the progress coefficient of the $(1, \lambda)$ strategy.

For an empirical investigation of the self-adaptation mechanism defined by the mutation operator variants (i)–(iii), Schwefel (1987, 1989, 1992) used the following three objective functions which are specifically tailored to the number of learnable strategy parameters in these cases.

(i)   Function

$$f_1(\boldsymbol{x}) = \sum_{i=1}^{n} x_i^2 \qquad (21.13)$$

requires learning of one common standard deviation σ, i.e. $n_\sigma = 1$.

(ii)   Function

$$f_2(\boldsymbol{x}) = \sum_{i=1}^{n} i x_i^2 \qquad (21.14)$$

requires learning of a suitable *scaling* of the variables, i.e. $n_\sigma = n$.

(iii)   Function

$$f_3(\boldsymbol{x}) = \sum_{i=1}^{n} \left(\sum_{j=1}^{i} x_j \right)^2 \qquad (21.15)$$

requires learning of a positive definite *metrics*, i.e. individual σ_i and $n_\alpha = n(n-1)/2$ different covariances.

As a first experiment, Schwefel compared the convergence velocity of a $(1, 10)$ and a $(1 + 10)$ evolution strategy with $n_\sigma = 1$ on the sphere model f_1 with $n = 30$. The results of a comparable experiment performed by the present author (averaged over ten independent runs, with the standard deviations initialized with a value of 0.3) are shown in figure 21.2 (top), where the convergence velocity or progress is measured by $\log((f_{\min}(0)/f_{\min}(g))^{1/2})$ with $f_{\min}(g)$ denoting the objective function value in generation g. It is somewhat counterintuitive to observe that the nonelitist $(1, 10)$ strategy, where all offspring individuals might be worse than the single parent, performs *better* than the elitist $(1 + 10)$ strategy. This can be explained, however, by taking into account that the self-adaptation of standard deviations might generate an individual with a good objective function value but an inappropriate value of σ for the next generation. In the case of a plus strategy, this inappropriate standard deviation might survive for a number of generations, thus hindering the combined process of search and adaptation. The resulting periods of stagnation can be prevented by allowing the good search point to be forgotten, together with its inappropriate step size. From this experiment, Schwefel concluded that the nonelitist (μ, λ) selection mechanism is an important condition for a successful self-adaptation of strategy parameters. Recent experimental findings by Gehlhaar and Fogel (1996) on objective functions more complicated than the sphere model give some evidence, however, that the elitist strategy performs as well as or even better than the (μ, λ) strategy in many practical cases.

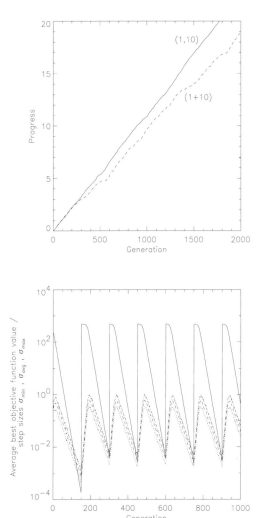

Figure 21.2. Top: a comparison of the convergence velocity of a $(1, 10)$ strategy and a $(1+10)$ strategy in the case of the sphere model f_1 with $n = 30$ and $n_\sigma = 1$. Bottom: the best objective function value and minimum, average, and maximum standard deviation in the population plotted over the generation number for the time-varying sphere model. The results were obtained by using a $(15, 100)$ evolution strategy with $n_\sigma = 1$, $n = 30$, without recombination.

For a further illustration of the self-adaptation principle in case of the sphere model f_1, we use a time-varying version where the optimum location $x^* = (x_1^*, \ldots, x_n^*)$ is changed every 150 generations. Ten independent experiments for $n = 30$ and 1000 generations per experiment are performed with a $(15, 100)$ evolution strategy (without recombination). The average best objective function value (solid curve) and the minimum, average, and maximum standard deviations $\sigma_{\min}$, σ_{avg}, and $\sigma_{\max}$ are shown in the lower part of figure 21.2. The curve of the objective function value clearly illustrates the linear convergence of the algorithm during the first search interval of 150 generations. After shifting the optimum location at generation 150, the search stagnates for a while at the bad new position before the linear convergence is observed again.

The behavior of the standard deviations, which are also plotted in figure 21.2 (bottom), clarifies the reason for the periods of stagnation of the objective function values: self-adaptation of standard deviations works both by decreasing them during the periods of linear convergence and by increasing them during the periods of stagnation, back to a magnitude such that they have an impact on the objective function value. This process of standard deviation increase, which occurs at the beginning of each interval, needs some time which does not yield any progress with respect to the objective function value. According to Beyer (1995b), the number of generations needed for this adaptation is inversely proportional to τ_0^2 (that is, proportional to n) in the case of a $(1, \lambda)$ evolution strategy.

In the case of the objective function f_2, each variable x_i is differently scaled by a factor $i^{1/2}$, such that self-adaptation requires the scaling of n different σ_i to be learned. The optimal settings of standard deviations $\sigma_i^* \propto i^{-1/2}$ are also known in advance for this function, such that self-adaptation can be compared to an evolution strategy using optimally adjusted σ_i for mutation. The result of this comparison is shown in figure 21.3 (left), where the convergence velocity is plotted for $(\mu, 100)$ evolution strategies as a function of μ, the number of parents, for both the self-adaptive strategy and the strategy using the optimal setting of σ_i.

It is not surprising to see that, for the strategy using optimal standard deviations σ_i, the convergence rate is maximized for $\mu = 1$, because this setting exploits the perfect knowledge in an optimal sense. In the case of the self-adaptive strategy, however, a clear maximum of the progress rate is reached for a value of $\mu = 12$, and both larger and smaller values of μ cause a strong loss of convergence speed. The collective performance of about 12 imperfect parents, achieved by means of self-adaptation, is almost equal to the performance of the perfect $(1, 100)$ strategy and outperforms the collection of 12 perfect individuals by far. This experiment indicates that self-adaptation is a mechanism that requires the existence of a knowledge diversity (or diversity of internal models), i.e. a number of parents larger than one, and benefits from the phenomenon of collective (rather than individual) intelligence.

Concerning the objective function f_3, figure 21.3 (bottom) shows a comparison of the progress for a $(15, 100)$ evolution strategy with $n_\sigma = n = 10$, $n_\alpha = 0$ (that is, no correlated mutations) and $n_\alpha = n(n - 1)/2 = 45$ (that is,

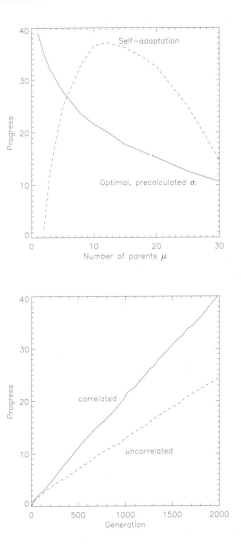

Figure 21.3. Top: the convergence velocity on f_2 for a $(\mu, 100)$ strategy with $\mu \in \{1, \ldots, 30\}$ for the self-adaptive evolution strategy and the strategy using optimum prefixed values of the standard deviations σ_i. Bottom: a comparison of the convergence velocity of a $(15, 100)$ strategy with correlated mutations in the case of the function f_3 with $n = n_\sigma = 10$, $n_\alpha = 45$ and with self-adaptation of standard deviations only (uncorrelated) for $n = n_\sigma = 10$, $n_\alpha = 0$.

full correlations). In both cases, intermediary recombination of object variables, global intermediary recombination of standard deviations, and no recombination of the rotation angles is chosen. The results demonstrate that, by introducing the covariances, it is possible to increase the effectiveness of the collective learning process in case of arbitrarily rotated coordinate systems. Rudolph (1992) has shown that an approximation of the Hessian matrix could be computed by correlated mutations with an upper bound of $\mu + \lambda = (n^2 + 3n + 4)/2$ on the population size, but the typical settings ($\mu = 15$, $\lambda = 100$) are often not sufficient to achieve this (an experimental investigation of the scaling behavior of correlated mutations with increasing population sizes and problem dimension has not yet been performed).

The choice of a logarithmic normal distribution for the modification of the standard deviations σ_i in connection with a multiplicative scheme in equations (21.4), (21.6), and (21.9) is motivated by the following heuristic arguments (see Schwefel 1977, p 168):

(i)   A multiplicative process preserves positive values.
(ii) The median should be equal to one to guarantee that, on average, a multiplication by a certain value occurs with the same probability as a multiplication by the reciprocal value (i.e. the process would be neutral under the absence of selection).
(iii) Small modifications should occur more often than large ones.

The effectiveness of this multiplicative logarithmic normal modification is currently also acknowledged in evolutionary programming, since extensive empirical investigations indicate some advantage of this scheme over the original additive self-adaptation mechanism used in evolutionary programming (Saravanan 1994, Saravanan and Fogel 1994, Saravanan *et al* 1995), where

$$\sigma_i' = \sigma_i(1 + \alpha N(0, 1)) \tag{21.16}$$

(with a setting of $\alpha \approx 0.2$ (Saravanan *et al* 1995)). Recent investigations indicate, however, that this becomes reversed when noisy objective functions are considered, where the additive mechanism seems to outperform multiplicative modifications (Angeline 1996).

The study by Gehlhaar and Fogel (1996) also indicates that the order of the modifications of x_i and σ_i has a strong impact on the effectiveness of self-adaptation: it is important to mutate the standard deviations first and to use the mutated standard deviations for the modification of object variables. As the authors point out in that study, the reversed mechanism might suffer from generating offspring that have useful object variable vectors but bad strategy parameter vectors, because these have not been used to determine the position of the offspring itself.

Concerning the sphere model f_1 and a $(1, \lambda)$ strategy, Beyer (1995b) has recently indicated that equation (21.16) is obtained from equation (21.4) by Taylor expansion breaking off after the linear term, such that both mutation mechanisms should behave identically for small settings of the learning rates τ_0 and α, when $\tau_0 = \alpha$. This was recently confirmed by Bäck and Schwefel (1996) with some experiments for the time-varying sphere model. Moreover,

Beyer (1995b) also shows that the self-adaptation principle works for a variety of different probability density functions for the modification of step sizes; that is, it is a very robust technique. For $n_\sigma = 1$, even the simple mutational step size control

$$\sigma' = \begin{cases} \sigma\alpha & \text{if } u \sim U(0, 1) \leq \frac{1}{2} \\ \sigma/\alpha & \text{if } u \sim U(0, 1) > \frac{1}{2} \end{cases} \tag{21.17}$$

of Rechenberg (1994, p 47) provides a reasonable choice. A value of $\alpha = 1.3$ of the learning rate is proposed by Rechenberg.

In concluding this subsection, recent approaches to substituting the normal distribution used for the modification of object variables x_i by other probability densities are worth mentioning. As outlined by Yao and Liu (1996), the one-dimensional Cauchy density function

$$f(x) = \frac{1}{\pi} \frac{t}{t^2 + (x - u)^2} \tag{21.18}$$

is a good candidate, because its shape resembles that of the Gaussian density function, but approaches the axis so slowly that expectation (and higher moments) do not exist. Consequently, it is natural to hope that the Cauchy density increases the probability of leaving local optima. Because the moments of a one-dimensional Cauchy distribution do not exist, Yao and Liu (1996) use the same self-adaptation mechanism as described by equations (21.6) and (21.7) with the only modification to substitute the standardized normally distributed $N(0, 1)$ in equation (21.7) by a random variable with one-dimensional Cauchy distribution with parameters $u = 0$, $t = 1$. A realization of such a random variable can be generated by dividing the realizations of two independent standard normally distributed random variables.

Using a large set of 23 test functions, the authors of this study conclude that their new algorithm (called *fast evolutionary programming*) performs better than the implementation using the normal rather than Cauchy distribution especially for multimodal functions with many local optima while being comparable to the normal distribution for unimodal and multimodal functions with only a few local optima.

Finally, the most general variant of self-adaptation seems to consist in the self-adaptation of the whole probability density function itself rather than having a fixed density and adapting one or more control parameters of that density. Davis (1994) implements this idea by representing a one-dimensional continuous probability density function by a discrete mutation histogram with 101 bars in width over a region of interest $[a, b]$. The heights of the histogram bars are integer values $h_0 h_1 \ldots h_{101}$, the histogram specifies a region that ranges from $x - (b - a)/2$ to $x + (b - a)/2$ around the current value x of a solution, and h_{50} is centered at the current value of the solution.

In Davis' implementation, an object variable is mutated by choosing a new solution value from the probability density function over the histogram's range. Afterwards, the mutation density is modified by choosing a histogram bar using the same probability density function, and incrementing the bar value with a probability proportional to the bar height. Using this mechanism for self-adaptation of the probability density function itself, Davis found that for landscapes with local optima the shape of the probability density function is adapted to reflect the landscape structure with respect to the location of the optima. Moreover, the formation of a peaked center in each of the mutation histograms is interpreted by Davis as a hint that the normal distribution naturally emerges as a good choice for a wide range of fitness landscapes.

While it is not surprising to see from these experiments that the structure of a one-dimensional objective function can be learned by self-adaptation, it is necessary here to emphasize that the extension of this approach to $n \gg 1$ dimensions fails because the discretized representation of an arbitrary multivariate distribution would require c^n histogram bars (c being the number of bars in one dimension). The condensed representation of a multivariate distribution by a few control parameters which are self-adapted (as in equations (21.9)–(21.11)) is a more appropriate method to handle the higher-dimensional case than the representation suggested by Davis.

21.2.2 Binary search spaces

A transfer of the self-adaptation principle from evolution strategies to the mutation probability $p_m \in [0, 1]$ of canonical (i.e. with a binary representation of individuals) genetic algorithms was first proposed by Bäck (1992b). Based on the binary representation $b = (b_1 \ldots b_\ell) \in \{0, 1\}^\ell$ of object variables (often, continuous object variables $x_i \in [u_i, v_i] \subset \mathbb{R}$, $i = 1, \ldots, n$, are represented in canonical genetic algorithms by binary strings and a Gray code; see Chapter 2 for details), Bäck extended the representation by an additional ℓ' (or $n\ell'$) bits to encode either one or n mutation probabilities (the latter can only be applied if the genotype b explicitly splits into n logical subparts encoding different object variables) as part of each individual. Because of the restricted applicability of the general case of n mutation probabilities, we discuss only the case of one mutation probability here.

An individual $a = (b, p)$ consists of the binary vector $b = (b_1 \ldots b_\ell) \in \{0, 1\}^\ell$ representing the object variables and the binary vector $p = (p_1 \ldots p_{\ell'}) \in \{0, 1\}^{\ell'}$ representing the individual's mutation rate p_m according to the decoding function $\Gamma_{0,1,\ell'}$, which is defined as follows:

$$\Gamma_{u,v,\ell}(b_1 \ldots b_\ell) = u + (v - u)\frac{\sum_{i=0}^{\ell-1}\left(\bigoplus_{j=1}^{i+1} b_j\right) 2^i}{2^\ell - 1}. \tag{21.19}$$

Here, $\oplus$ denotes summation modulo 2, such that a Gray code and a linear mapping of the decoded integer to the range $[u, v]$ are used. With the definition given in equation (21.19), p_m and p are related by $p_m = \Gamma_{0,1,\ell'}(p)$, and the mutation operator for self-adapting p_m proceeds by mutating p with mutation rate p_m, thus obtaining p' and $p'_m = \Gamma_{0,1,\ell'}(p')$, and then mutating b with mutation rate p'_m; that is,

$$p_m = \Gamma_{0,1,\ell'}(p_1 \ldots p_{\ell'}) \tag{21.20}$$

$$p'_i = \begin{cases} p_i & u > p_m \\ 1 - p_i & u \le p_m \end{cases} \tag{21.21}$$

$$p'_m = \Gamma_{0,1,\ell'}(p'_1 \ldots p'_{\ell'}) \tag{21.22}$$

$$b'_j = \begin{cases} b_j & u > p'_m \\ 1 - b_j & u \le p'_m. \end{cases} \tag{21.23}$$

As usual, $u \sim U([0,1))$ denotes a uniform random variable sampled anew for each $i \in \{1, \ldots, \ell'\}$ and $j \in \{1, \ldots, \ell\}$.

This mutation mechanism was experimentally tested on a few continuous, high-dimensional test functions (the sphere model, the weighted sphere model, and the generalized Rastrigin function) with a genetic algorithm using (μ, λ) selection, and outperformed a canonical genetic algorithm with respect to convergence velocity as well as convergence reliability (Bäck 1992b). Concerning the selection method, these experiments demonstrated that the (μ, λ) selection (with $\mu = 10$, $\lambda = 50$) clearly outperformed proportional selection and facilitated much larger average mutation rates in the population than proportional selection did (for proportional selection, mutation rates quickly dropped to an average value of 0.001, roughly a value of $1/\ell$, while for $(10, 50)$ selection mutation rates as large as 0.005 were maintained by the algorithm).

Based on Bäck's work, Smith and Fogarty (1996) recently incorporated the self-adaptation method described by equations (21.20)–(21.23) into a steady-state (or $(\mu + 1)$, in the terminology of evolution strategies) genetic algorithm, where just one new individual is created and substituted within the population at each cycle of the main loop of the algorithm. The new individual is generated by recombination (parents are chosen according to proportional selection), followed by an internal $(1, c)$ strategy which generates c mutants according to the self-adaptive method described above and selects one of them (either deterministically as usual, or according to proportional selection) as the offspring of the $(\mu + 1)$ strategy. The authors investigated several policies for deleting an individual from the population (deletion of the worst, deletion of the oldest), recombination operators (which, however, had no clear impact on the results at all), mutation encoding (standard binary code, Gray code, exponential code), and the value of $c \in \{1, 2, 5, 10\}$ on NK landscapes with $N = 16$ and $K \in \{0, 4, 8, 15\}$. From these experiments, Smith and Fogarty derived a number of important conclusions regarding the best policies for the self-adaptation mechanism, namely:

(i) Replacing the oldest of the population with the best offspring, conditional on the latter being the better of the two, is the best selection and deletion policy. Because replacing the oldest (rather than the worst) drops the elitist property of the $(\mu + 1)$ strategy, this confirms observations from evolution strategies that self-adaptation needs a nonelitist selection strategy to work successfully (see above).

(ii) A value of $c = 5$ was consistently found to produce best results, such that the necessity to produce a surplus of offspring individuals as found by Bäck (1992b) and the $1/5$ success rule are both confirmed.

(iii) Gray coding and standard binary coding showed similar performance, both substantially outperforming the exponential encoding. On the most complex landscapes, however, the Gray coding also outperformed standard binary coding.

The comparison of the self-adaptive mutation mechanism with all standard fixed mutation rate settings (see Chapter 18 for an overview) clarified the general advantage of self-adaptation by significantly outperforming these fixed mutation rate settings.

The method described so far for self-adapting the mutation rates in canonical genetic algorithms was historically developed on the basis of the assumption that both the object variables and the strategy parameters should be represented by binary strings. It is clear from research on evolution strategies and evolutionary programming, however, that it should also be possible to incorporate the mutation rate $p_m \in [0, 1]$ directly into the genotype of individuals $a = (b, p_m) \in \{0, 1\}^\ell \times [0, 1]$ and to formulate a mutation operator that mutates p_m rather than its binary representation. Recently, Bäck and Schütz (1995, 1996) proposed a first version of such a self-adaptation mechanism, which was successfully tested for the mixed-integer problem of optimizing optical multilayer systems as well as for a number of combinatorial optimization problems with binary object variables.

Based on a number of requirements similar to those formulated by Schwefel for evolution strategies, namely that

(i)   the expected change of p_m by repeated mutations should be equal to zero,
(ii) mutation of $p_m \in]0, 1[$ must yield a feasible mutation rate $p'_m \in]0, 1[$,
(iii) small changes should be more likely than large ones, and
(iv) the median should equal one,

these authors proposed a mutation operator of the form

$$p'_m = \left(1 + \frac{1 - p_m}{p_m} \exp(-\gamma N(0, 1))\right)^{-1} \tag{21.24}$$

$$b'_j = \begin{cases} b_j & u > p'_m \\ 1 - b_j & u \le p'_m \end{cases} \tag{21.25}$$

where γ ($\gamma = 0.2$ was chosen in the experiments of the authors) is a learning rate analogue to τ_0 in equation (21.4). A direct comparison of this operator with

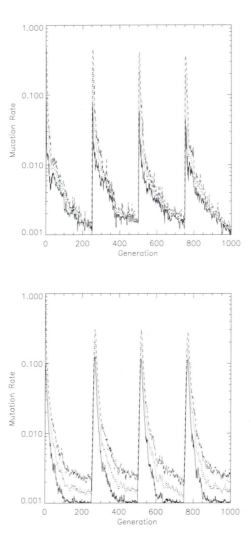

Figure 21.4. The self-adaptation of mutation rates is shown here for the time-varying counting ones function with $\ell = 1000$ and $(15, 100)$ selection, without recombination. The upper plot shows the minimum, average, and maximum mutation rates that occur in the population when the binary representation of p_m is used, while the lower plot shows the corresponding mutation rates when the mutation operator according to (21.24) is used.

the one using a binary representation of p_m has not yet been performed, but it has already been observed by Bäck and Schütz (1996) that the learning rate γ is a critical parameter of equation (21.24), because it determines the velocity of the self-adaptation process. In contrast to this, the method described by equations (21.20)–(21.23) eliminates the mutation rate completely as a parameter of the algorithm, such that it provides a more robust algorithm.

In analogy with the experiment performed with evolution strategies on the time-varying version of the sphere model, the self-adaptation mechanisms for mutation rates are tested with a time-varying version of the binary counting ones problem $f(b) = \sum_{i=1}^{\ell} b_i \rightarrow$ min, which is modified by switching between f and $f'(b) = \ell - f(b)$ every g generations. The experiment is performed for $\ell = 1000$ and $g = 250$ with a self-adaptive genetic algorithm using (15, 100) selection, but without crossover. The results for the minimum, average, and maximum mutation rates are shown in figure 21.4 for the mutation mechanism according to equations (21.20)–(21.23) (upper plot) and for the mutation mechanism according to equation (21.24) (lower plot). It is clear from these figures that both mutation schemes facilitate the necessary adaptation of mutation rates, following a near-optimal schedule that exponentially decreases from large mutation rates ($p_m \approx 0.5$) at the beginning of the search to mutation rates of the order of $1/\ell$ in the final stage of the search. This behavior is in perfect agreement with the theoretical knowledge about the optimal mutation rate for the counting ones function (see e.g. Chapter 18 or the article by Bäck (1993)), but the available diversity of mutation rates in the population is smaller when the binary representation (upper plot) is used than with the continuous representation of p_m.

The corresponding best objective function values in the population are shown in figure 21.5, and give clear evidence that the principle works well for both of the self-adaptation mechanisms presented here—thus again confirming the result of Beyer that the precise form of the probability density function used for modifying the mutation rates does not matter very much. Also in binary search spaces, self-adaptation is a powerful and robust technique, and the following sections demonstrate that this is true for other search spaces as well.

21.2.3 Integer search spaces

For the general integer programming problem

$$\max\{f(x) \mid x \in M \subseteq \mathbb{Z}^n\} \tag{21.26}$$

where $\mathbb{Z}$ denotes the set of integers, Rudolph (1994) presented an algorithm that self-adapts the total step size s of the variation of x. By applying the principle of maximum entropy, i.e. the search for a distribution that is spread out as uniformly as possible without contradicting the given information, he demonstrated that a multivariate, symmetric extension of the geometric distribution is most suitable for the distribution of the variation $k = (k_1, \ldots, k_n)^{\mathrm{T}}$ ($k_i \in \mathbb{Z}$) of the object variable vector x. For a multivariate random variable Z, which is distributed

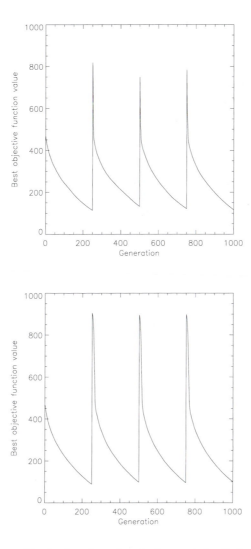

Figure 21.5. The best objective function values in the population for both self-adaptation mechanisms shown in figure 21.4.

according to the probability density function

$$P\{Z_1 = k_1, \ldots, Z_n = k_n\} = \prod_{i=1}^{n} P\{Z_i = k_i\} \qquad (21.27)$$

$$= \left(\frac{p}{2-p}\right)^n \prod_{i=1}^{n} (1-p)^{|k_i|} \qquad (21.28)$$

$$= \left(\frac{p}{2-p}\right)^n (1-p)^{\|k\|_1} \qquad (21.29)$$

where $\|k\|_1 = \sum_{i=1}^{n} |k_i|$ denotes the ℓ_1 norm, the expectation of the ℓ_1 norm of Z (i.e. the mean step size s) is given by

$$s = \mathbf{E}(\|Z\|_1) = n\frac{2(1-p)}{p(2-p)}. \qquad (21.30)$$

In full analogy with equation (21.4), the mutation operator proposed by Rudolph modifies step size s and object variables x_i according to

$$s' = s\exp(\tau_0 N(0,1)) \qquad (21.31)$$
$$x_i' = x_i + G_i(p') \qquad (21.32)$$

where $G_i(p')$ denotes a realization of a one-dimensional random variable with probability density function

$$P\{G_i = k\} = \frac{p'}{2-p'}(1-p')^{|k|} \qquad (21.33)$$

i.e. the symmetric generalization of the geometric distribution with parameter p', and p' is obtained from s' according to equation (21.30) as follows:

$$p' = 1 - \frac{s'/n}{1 + (1 + (s'/n)^2)^{1/2}}. \qquad (21.34)$$

Algorithmically, a realization $G_i(p')$ can be generated as the difference of two independent geometrically distributed random variables (both with parameter p'). A geometric random variable G is obtained from a uniform random variable U over $[0,1) \subset \mathbb{R}$ according to the transformation

$$G = \left\lfloor \frac{\log(1-U)}{\log(1-p')} \right\rfloor. \qquad (21.35)$$

Except for the integer representation and the mutation operator, the algorithm developed by Rudolph is based on an ordinary evolution strategy with (μ, λ) selection (with $\mu = 30$, $\lambda = 100$ for the experimental runs), intermediary recombination of the step sizes s, and no recombination of the object variable vector (i.e. when two individuals are recombined, their step sizes are averaged and the object variable vector is chosen from the first or second parent at random). The algorithm was empirically tested by Rudolph (1994) on five nonlinear integer programming problems and located the global optima of these problems for at least 80 percent of 1000 independent runs per problem.

21.2.4 Finite-state machines

Methods to apply the self-adaptation principle to finite-state machine representations as used by evolutionary programming for sequence prediction tasks have recently been presented by Fogel *et al* (1994, 1995). These authors discuss two different methods for self-adaptation of mutability parameters (i.e. the probabilities of performing any of the possible mutations of state addition, state deletion, changes of output symbols, the initial state, and a next-state transition) associated with each component of a finite-state machine. The mutability parameters p_i are all initially set to a minimum value of 0.001 and mutated according to a historically older version of equation (21.16) (Fogel *et al* 1991):

$$p_i' = p_i + \alpha N(0, 1) \tag{21.36}$$

where $\alpha = 0.01$.

The methods are different with respect to the selection of a machine component for mutation. In *selective self-adaptation*, a component is selected for each type of mutation on the basis of relative selection probabilities $p_i / \sum p_k$, where p_i is the mutability parameter for the ith component and the summation is taken with k running over all components. For this scheme, separate mutability parameters are maintained for each state, each output symbol, and each next-state transition.

In contrast to this, *multimutational self-adaptation* denotes a mechanism where each mutability parameter designates the absolute probability of modification for that particular component, such that the probability for each component to be mutated is independent of the probabilities of other components. Consequently, multimutational self-adaptation is expected to offer greater diversity in the types of offspring machine that could be generated from a parent, and the learning rate α of the approach is extremely important under this approach ($0.005 \leq \alpha \leq 0.999$ was generally enforced in the experiments).

Both approaches were tested by the authors on two simple prediction tasks and consistently outperformed the evolutionary programming method without self-adaptation. While on the simpler problem the selective self-adaptation method had a slightly better performance than the multimutational method, the latter performed better on the more complex problem, thus indicating the need to explore a larger diversity by the mutation operator in this case. Certainly, more work is needed to assess the strengths and weaknesses of both approaches for self-adaptation on finite-state machines, but once again the robustness and wide applicability of this general principle for on-line learning of mutational control parameters has been demonstrated by these experiments.

21.3 Recombination operators

In contrast to the mutation operator, recombination in evolutionary algorithms has received much less attention for self-adaptation of the operator's characteristics (e.g. the number and location of crossover points) and parameters (e.g. the application probability or segment exchange probability). This can be explained in part by the historical emergence of the algorithmic principle in connection with the mutation operator in evolution strategies and evolutionary programming, but it might also be argued that the implicit link between strategy parameters and their impact on the fitness of an individual is not as tight for recombination as the self-adaptation idea requires (this argument is supported by recent empirical findings indicating that the set of problems where crossover is useful might be smaller than expected (Eshelman and Schaffer 1993)). Research concerning the self-adaptation of recombination operators is still in its infancy, and the two examples reported here for binary search spaces can only give an impression of the many open questions that need to be answered in the future.

21.3.1 Binary search spaces

For canonical genetic algorithms and a multipoint crossover operator, Schaffer and Morishima (1987) developed a method for self-adaptation of both the number and location of crossover points. These strategy parameters (i.e. the crossover points) are incorporated in an individual by attaching to the end of each object variable vector $b = (b_1 \ldots b_\ell) \in \{0, 1\}^\ell$ another binary vector c of the same length. The bits of this strategy parameter vector are interpreted as *crossover punctuations*, where a one at position i in vector c indicates that a crossover point occurs at the corresponding position i in the object variable vector. During initialization, the probability of generating a one in the vector c is set to a rather small value $p_{xo} = 0.04$.

Crossover between two strings then works by copying the bits from each parent string one by one to one of the offspring from left to right, and when a punctuation mark is encountered in either parent, the bits begin going to the other offspring. The punctuation marks themselves are also passed on to the offspring, when this happens, such that the strategy parameter vector is also subject to recombination. Furthermore, the usual mutation by bit inversion is also applied to the strategy parameter vector, such that the principles of self-adaptation are fully applied in the method proposed by Schaffer and Morishima (1987).

For an empirical investigation of the *punctuated crossover* method, the authors used the five test functions proposed by De Jong (1975) and the on-line performance measure (i.e. an average of all trials in a run). The self-adaptive crossover operator was found to statistically outperform a canonical genetic algorithm on four of the five functions while being no worse on the other. Looking at the dynamics of the distribution of punctuation marks over time, the authors observed that some loci tend to accumulate more punctuation marks than others as time progresses and the locations of these concentrations change over time. Concerning the dynamics of the average number of crossover events that occur per mating, the authors found that the number of 'productive' crossover events (i.e. those events where nonidentical gene segments are swapped between parents and offspring different from the parents is produced) remained nearly constant and correlated strongly with ℓ (and implicitly with n, the dimension of the real-valued search space the binary strings are mapped to in this parameter optimization task). These results are certainly interesting and deserve a more detailed investigation, especially under experimental conditions where the effect of a recombination operator is well understood and the optimal operator to be encountered by self-adaptation is known in advance. The sphere model with continuous representation of variables could be a good candidate for such an experiment, because the theory of discrete and intermediary recombination is relatively well understood for this function (Beyer 1995a).

As clarified by Spears (1995), it is not clear whether the success of Schaffer and Morishima's punctuated crossover is due to the self-adaptation of crossover points or whether it stems from the simple fact that they compared a canonical genetic algorithm with one-point crossover with the self-adaptive method that, on average, was using n-point crossover with $n > 1$. Spears (1995) investigated a simple method called one-bit self-adaptation, where a single strategy parameter bit added to an individual indicated whether uniform crossover or two-point crossover was performed on the parents (ties are broken randomly). Experiments were performed on the so-called *N-peak problems*, in which each problem has one optimal solution and $N - 1$ suboptimal solutions. The cases $N \in \{1, 6\}$ were investigated for 30 bits and 900 bits, where the latter problem contains 870 dummy bits that do not change the objective function, but have some impact on the effectiveness of the crossover operators.

The author performed a control experiment to determine the best crossover operator on these problems, and then ran the self-adaptation method on the problems. Concerning the best-so-far curves, the performance of self-adaptation consistently approached the performance of the best crossover operator alone, but in the case of the six-peak 900-bit problem, the dominant operator chosen by self-adaptation (two-point crossover in more than 90% of all cases) was *not* the operator identified in the control experiment (uniform crossover). As an overall conclusion of further experiments along this line, Spears (1995) clarified that the key feature of the 'self-adaptation' method used here is not to provide the possibility of adapting towards choosing the best operator, but rather the diversity provided to the algorithm by giving it access to two different recombination operators for exploitation during the search. Consequently, it might be worthwhile to run an evolutionary algorithm with a larger set of search operators than is customary, even if the algorithm is not self-adaptive, and the aspect of an additional benefit caused by the available diversity of strategy parameters is emphasized again by this observation.

21.4   Conclusions

Numerous approaches for the self-adaptation of strategy parameters within evolutionary algorithms as discussed in the previous sections clarify that the fundamental underlying principle of self-adaptation—evolving both the object variables *and* the strategy parameters simultaneously—works under a variety of conditions regarding the search space and the variation mechanism for strategy parameters by exploiting the implicit link between strategy parameters (or *internal models*) of the individuals and their fitness. This seems to work best for the unary mutation operator, because the strategy parameters have a direct impact on the object variables of a single individual and are either selected for survival and reproduction or discarded, depending on whether their impact is beneficial or detrimental.

This robustness of the method clearly indicates that a fundamental principle of evolutionary processes is utilized here, and in fact it is worth mentioning that the base pair mutation rate of mammalian organisms is in part regulated by its own genotype by *repair enzymes* and *mutator genes* encoded on the DNA. The former are able to repair a variety of damage of the genome, while the latter increase the mutation rate of other parts of the genome (see Gottschalk 1989, pp 269–71, 182).

Concerning its relevance as a parameter control mechanism in evolutionary algorithms, self-adaptation clearly has the advantage of reducing the number of exogenous control parameters of these algorithms and thereby releasing the user from the costly process of fine tuning these parameters by hand. Moreover, it is well known that constant control parameter settings (e.g. of the mutation rate in canonical genetic algorithms) are far from being optimal and that self-adaptation principles are able to generate a nearly optimal dynamic schedule of these parameters (see e.g. the article by Bäck (1992a) and the examples for the time-varying sphere model and counting ones function as presented in Section 21.2).

While the principle is of much practical usefulness and has demonstrated its power and robustness in many examples, many open questions remain to be answered. The most important questions are those for the optimal conditions for self-adaptation concerning the choice of a selection operator, population sizes, and the probability density function used for strategy parameter modifications, i.e. for the algorithmic circumstances required. This also raises the question for an appropriate optimality criterion for self-adaptation, having in mind that maximizing speed of adaptation might be good for holding an optimum within dynamically changing environments rather than for emphasizing global convergence properties. The speed of adaptation is controlled in some of the self-adaptation approaches by exogenous learning rates (e.g. τ_0 in equation (21.4), τ and τ' in equation (21.6), and α in equation (21.16)), and the 'optimal' setting of these learning rates usually emphasizes convergence velocity rather than global convergence of the evolutionary algorithm, such that for multimodal objective functions a different setting of the learning rates, implying slower adaptation, might be more appropriate. Finally, it is important to recognize that the term self-adaptation is used to characterize a wide spectrum of different possible behaviors, ranging from the precise learning of the time-dependent optimal setting of a single control parameter (such as σ for the time-varying sphere model) to the creation of a diversity of different strategy parameter values which are available in the population for utilization by the individuals.

It has always been emphasized by Schwefel (1987, 1989, 1992) that diversity of the internal models is a key ingredient to the synergistic effect of self-adaptation, facilitating a collection of individuals equipped with imperfect, diverse internal models of their environment to perform collectively as well as or even better than a single expert individual with precise, full knowledge of the environment does. While some of the implementations of self-adaptation certainly exploit more the diversity of parameter settings rather than adapting them, the key to the success of self-adaptation seems to consist in using both sides of the coin to facilitate reasonably fast adaptation (and, as a consequence, a good convergence velocity) and reasonably large diversity (and a good convergence reliability) at the same time.

References

Angeline P J 1996 The effects of noise on self-adaptive evolutionary optimization *Proc. 5th Ann. Conf. on Evolutionary Programming* ed L J Fogel, P J Angeline and T Bäck (Cambridge, MA: MIT Press)

Arabas J, Michalewicz Z and Mulawka J 1994 GAVaPS—a genetic algorithm with varying population size *Proc. 1st IEEE Conf. on Evolutionary Computation (Orlando, FL, June 1994)* (Piscataway, NJ: IEEE) pp 73–8

Bäck T 1992a The interaction of mutation rate, selection, and self-adaptation within a genetic algorithm *Parallel Problem Solving from Nature, 2 (Proc. 2nd Int. Conf. on Parallel Problem Solving from Nature, Brussels, 1992)* ed R Männer and B Manderick (Amsterdam: Elsevier) pp 85–94

——1992b Self-adaptation in genetic algorithms *Proc. 1st Eur. Conf. on Artificial Life* ed F J Varela and P Bourgine (Cambridge, MA: MIT Press) pp 263–71

——1993 Optimal mutation rates in genetic search *Proc. 5th Int. Conf. on Genetic Algorithms (Urbana-Champaign, IL, July 1993)* ed S Forrest (San Mateo, CA: Morgan Kaufmann) pp 2–8

Bäck T and Schütz M 1995 Evolution strategies for mixed-integer optimization of optical multilayer systems *Proc. 4th Ann. Conf. on Evolutionary Programming (San Diego, CA, March 1995)* ed J R McDonnell, R G Reynolds and D B Fogel (Cambridge, MA: MIT Press) pp 33–51

——1996 Intelligent mutation rate control in canonical genetic algorithms *Foundations of Intelligent Systems, 9th Int. Symp., ISMIS '96 (Lecture Notes in Artificial Intelligence 1079)* ed Z W Ras and M Michalewicz (Berlin: Springer) pp 158–67

Bäck T and Schwefel H-P 1996 Evolutionary computation: an overview *Proc. 3rd IEEE Conf. on Evolutionary Computation* (Piscataway, NJ: IEEE) pp 20–9

Beyer H-G 1995a Toward a theory of evolution strategies: on the benefits of sex—the $(\mu/\mu, \lambda)$-theory *Evolutionary Comput.* **3** 81–111

——1995b Toward a theory of evolution strategies: self-adaptation *Evolutionary Comput.* **3** 311–48

Davis L 1989 Adapting operator probabilities in genetic algorithms *Proc. 3rd Int. Conf. on Genetic Algorithms (Fairfax, VA, June 1989)* ed J D Schaffer (San Mateo, CA: Morgan Kaufmann) pp 61–9

Davis M W 1994 The natural formation of Gaussian mutation strategies in evolutionary programming *Proc. 3rd Ann. Conf. on Evolutionary Programming (San Diego, CA, February 1994)* ed A V Sebald and L J Fogel (Singapore: World Scientific) pp 242–52

De Jong K A 1975 *An Analysis of the Behaviour of a Class of Genetic Adaptive Systems* PhD Thesis, University of Michigan

Eiben A E and Michalewicz Z 1996 Personal communication

Eshelman L J and Schaffer J D 1993 Crossover's niche *Proc. 5th Int. Conf. on Genetic Algorithms (Urbana-Champaign, IL, July 1993)* ed S Forrest (San Mateo, CA: Morgan Kaufmann) pp 9–14

Fogel D B 1992 *Evolving Artificial Intelligence* PhD Thesis, University of California

Fogel D B, Fogel L J and Atmar W 1991 Meta-evolutionary programming *Proc. 25th Asilomar Conf. on Signals, Systems and Computers (Pacific Grove, CA)* ed R R Chen, pp 540–5

Fogel L J, Angeline P J and Fogel D B 1995 An evolutionary programming approach to self-adaptation on finite state machines *Proc. 4th Ann. Conf. on Evolutionary Programming (San Diego, CA, March 1995)* ed J R McDonnell, R G Reynolds and D B Fogel (Cambridge, MA: MIT Press) pp 355–65

Fogel L J, Fogel D B and Angeline P J 1994 A preliminary investigation on extending evolutionary programming to include self-adaptation on finite state machines *Informatica* **18** 387–98

Gehlhaar D K and Fogel D B 1996 Tuning evolutionary programming for conformationally flexible molecular docking *Proc. 5th Ann. Conf. on Evolutionary Programming* ed L J Fogel, P J Angeline and T Bäck (Cambridge, MA: MIT Press)

Gottschalk W 1989 *Allgemeine Genetik* (Stuttgart: Thieme)

Rechenberg I 1994 *Evolutionsstrategie '94 (Werkstatt Bionik und Evolutionstechnik 1)* (Stuttgart: Frommann–Holzboog)

Rudolph G 1992 On correlated mutations in evolution strategies *Parallel Problem Solving from Nature, 2 (Proc. 2nd Int. Conf. on Parallel Problem Solving from Nature, Brussels, 1992)* ed R Männer and B Manderick (Amsterdam: Elsevier) pp 105–14

——1994 An evolutionary algorithm for integer programming *Parallel Problem Solving from Nature—PPSN III (Proc. Int. Conf. on Evolutionary Computation and 3rd Conf. on Parallel Problem Solving from Nature, Jerusalem, October 1994) (Lecture Notes in Computer Science 866)* ed Yu Davidor, H-P Schwefel and R Männer (Berlin: Springer) pp 139–48

Saravanan N 1994 Learning of strategy parameters in evolutionary programming: an empirical study *Proc. 3rd Ann. Conf. on Evolutionary Programming (San Diego, CA, February 1994)* ed A V Sebald and L J Fogel (Singapore: World Scientific)

Saravanan N and Fogel D B 1994 Evolving neurocontrollers using evolutionary programming *Proc. 1st IEEE Int. Conf. on Evolutionary Computation (Orlando, FL, June 1994)* vol 1 (Piscataway, NJ: IEEE) pp 217–22

Saravanan N, Fogel D B and Nelson K M 1995 A comparison of methods for self-adaptation in evolutionary algorithms *BioSystems* **36** 157–66

Schaffer J D and Morishima A 1987 An adaptive crossover distribution mechanism for genetic algorithms *Proc. 2nd Int. Conf. on Genetic Algorithms (Cambridge, MA, July 1987)* ed J J Grefenstette (Hillsdale, NJ: Erlbaum) pp 36–40

Schlierkamp-Voosen D and Mühlenbein H 1996 Adaptation of population sizes by competing subpopulations *Proc. 3rd IEEE Conf. on Evolutionary Computation* (Piscataway, NJ: IEEE) pp 330–5

Schwefel H-P 1977 *Numerische Optimierung von Computer-Modellen mittels der Evolutionsstrategie (Interdisciplinary Systems Research 26)* (Basel: Birkhäuser)

——1987 Collective intelligence in evolving systems *Ecodynamics, Contributions to Theoretical Ecology* ed W Wolff, C-J Soeder and F R Drepper (Berlin: Springer) pp 95–100

——1989 Simulation evolutionärer Lernprozesse *Erwin–Riesch Workshop on Systems Analysis of Biomedical Processes, Proc. 3rd Ebernburger Working Conf. of ASIM/GI* ed D P F Möller (Braunschweig: Vieweg) pp 17–30

——1992 Imitating evolution: collective, two-level learning processes *Explaining Process and Change—Approaches to Evolutionary Economics* ed U Witt (Ann Arbor, MI: University of Michigan Press) pp 49–63

Smith J and Fogarty T C 1996 Self adaptation of mutation rates in a steady state genetic algorithm *Proc. 3rd IEEE Conf. on Evolutionary Computation* (Piscataway, NJ: IEEE) pp 318–23

Spears W M 1995 Adapting crossover in evolutionary algorithms *Proc. 4th Ann. Conf. on Evolutionary Programming (San Diego, CA, March 1995)* ed J R McDonnell, R G Reynolds and D B Fogel (Cambridge, MA: MIT Press) pp 367–84

Yao X and Liu Y 1996 Fast evolutionary programming *Proc. 5th Ann. Conf. on Evolutionary Programming* ed L J Fogel, P J Angeline and T Bäck (Cambridge, MA: MIT Press)

22

Meta-evolutionary approaches

Bernd Freisleben

22.1 Working mechanism

After having defined the individuals of a population for a given problem, the designer of an evolutionary algorithm (EA)† is faced with the problem of deciding what types of operator and control parameter settings are likely to produce the best results. For example, the decisions to be made might include choices among the different variants of the selection, *crossover*, and *mutation* operators which have been suggested in the literature (Bäck 1996, Goldberg 1989a, Michalewicz 1994), and, depending on the operators, values for the corresponding control parameters such as the crossover probability, the mutation probability, and the population size (Booker 1987, De Jong 1975, Goldberg 1989b, Hesser and Männer 1990, Mühlenbein and Schlierkamp-Voosen 1995, Schaffer *et al* 1989) must be determined. The decision may be based on:

- systematically checking a range of operators and/or parameter values and assessing the performance of the EA (De Jong 1975, Schaffer *et al* 1989)
- the experiences reported in the literature describing similar application scenarios (Goldberg 1989a, Jog *et al* 1989, Oliver *et al* 1987, Starkweather *et al* 1991)
- the results of theoretical analyses for determining the optimal parameter settings (Goldberg 1989b, Hesser and Männer 1990, Nakano *et al* 1994).

Since these proposals are typically not universally valid, and therefore it may be that none of them produces satisfactory results for the particular problem considered, a more promising approach is to consider the search for the 'best' EA for a given problem as an optimization problem itself. This *metaoptimization* problem can then be solved by any of the general purpose optimization methods proposed in the literature, including well-known heuristic methods (Reeves 1993) such as simulated annealing (van Laarhoven and Aarts 1987) or tabu

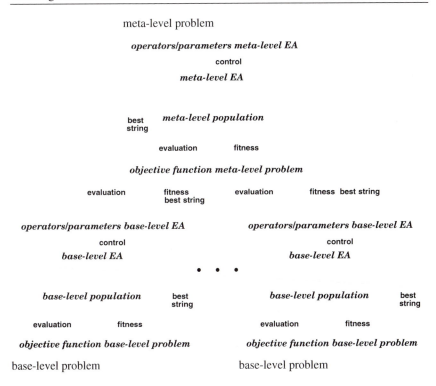

Figure 22.1. A meta-evolutionary approach.

search (Glover 1990)—but also by any type of evolutionary algorithm, leading to a *meta-evolutionary* approach.

In order to realize a meta-evolutionary approach, a two-level optimization strategy is required, as shown in figure 22.1. At the top level, a *metalevel* EA operates on a population of *base-level* EAs, each of which is represented by a separate individual. At the bottom level, the base-level EAs work on a population of individuals which represent possible solutions of the problem to be solved. Each of the base-level EAs runs independently to produce a solution of the problem considered, and the fitness of the solution influences the operation of the metalevel EA; the fitness of an individual representing a base-level EA is taken to be the fitness of the best solution found by the base-level EA in the entire run using the current parameters. The numbers of generations created on the two levels are independent of each other. The individual with the highest fitness ever found is expected to be the best EA for the original problem.

† The term *evolutionary algorithm* is used to denote any method of evolutionary computation, such as genetic algorithms (GA), evolution strategies (ESs), evolutionary programming (EP), and genetic programming (GP).

The information that needs to be represented to encode a base-level EA as a member of the metalevel population depends on the nature of the base-level problem. In general, since an EA is characterized by its operators and the parameter values to control them, there are essentially two kinds of information that need to be represented in the individuals of the metalevel population: first the particular variant of an evolutionary operator, and second the parameter values required for the selected variant. In some sense, this is a kind of variable-dimensional structure optimization problem.

The above description is implicitly based on the classical (steady-state) model of evolutionary computation, in which the operators and operator probabilities are specified before the start of an EA and remain unchanged during its run. However, several proposals have been made to allow the adaptation of the operator probabilities as the run progresses, e.g. in order to facilitate schema identification and reduce schema disruption (Davis 1991, Ostermeier *et al* 1994, White and Oppacher 1994). Furthermore, there is some empirical evidence that the most effective operator variants also vary during the search process (Davis 1991, Michalewicz 1994). Since in a meta-evolutionary approach the overall behavior of a base-level EA is typically used to evaluate its fitness in the metalevel population, the concept of self- *adaptation* (i.e. the evolution of strategy parameters on-line during the search, see Bäck 1996) is more suitable to support dynamically adaptive models of evolutionary computation. However, an advanced version of a meta-evolutionary approach might nevertheless be beneficial to analyze the impact of particular operator variants and parameter values on the solution quality during the different stages of the evolution process (Freisleben and Härtfelder 1993a, b) and might therefore provide valuable hints for dynamical operator/parameter adaptation.

22.2 Formal description

Let B be a base-level problem, I_B its solution space, $x = (x_1, \ldots, x_m) \in I_B$ a member of the solution space, and $F_B : I_B \to \mathbb{R}$ the objective function to be optimized (without loss of generality, we assume that F_B should be maximized, i.e. we are looking for an x^* such that $F_B(x^*) \geq F_B(x) \quad \forall x \in I_B$).

Furthermore, let $\text{EA}_{[V,F,I]} : V \to I$ be a generic evolutionary algorithm parameterized by the space V of vectors representing all possible settings of operator variants and parameters, a fitness function F, and the space I of individuals, which for each vector $v \in V$ returns an individual $x_v = \text{EA}_{[V,F,I]}(v) \in I$ with the best possible fitness.

Based on this generic definition, let us consider an evolutionary algorithm $EA_{[V_B,F_B,I_B]} : V_B \to I_B$ for problem B. In order to find a good solution of problem B, our goal is to find a parameter setting $v_B^* \in V_B$ such that

$$F_B(\text{EA}_{[V_B,F_B,I_B]}(v_B^*)) \geq F_B(\text{EA}_{[V_B,F_B,I_B]}(v_B)) \qquad \forall v_B \in V_B. \qquad (22.1)$$

Thus, the search for a good solution for the base-level problem B reduces to a metalevel problem M of maximizing the objective function $F_M : V_B \rightarrow \mathbb{R}$,

$$F_M(v_B) = F_B(\text{EA}_{[V_B, F_B, I_B]}(v_B)). \tag{22.2}$$

The definition of the objective function (22.2) allows us to treat the search for the best evolutionary algorithm for problem B as an optimization problem which may be solved by a meta-evolutionary algorithm $\text{EA}_{[V_M, F_M, V_B]} : V_M \rightarrow V_B$, where V_M is the space of all possible operator and parameter settings for the metalevel problem M. The aim of the meta-evolutionary algorithm is to find optimal parameter values

$$v_B^* = \text{EA}_{[V_M, F_M, V_B]}(v_M) \tag{22.3}$$

where v_M is the parameter setting useed for the meta-evolutionary algorithm.

22.3 Pseudocode

The pseudocode for an EA which is used to realize a meta-evolutionary approach is presented below. The code is based on the selection, recombination, mutation and replacement sequence of operations typically found in the genetic algorithm paradigm (see Sections B1.1 and B1.2 for a more detailed description); in this particular version, replacement of old individuals by new ones (line 8) is performed outside the **for**-loop (line 5); that is, the new population is created after all offsprings of a generation have been produced. Note that in order to distinguish between the base- and the metalevel EA, two new parameters (F, I) have been introduced in addition to the ones given in the pseudocode presented in Sections B1.1 and B1.2.

```
Input: μ, λ, t_max, F, I
Output: x* ∈ I, the best individual ever found.
procedure EA;
1   t ← 0;
2   P(t) ← initialize(μ, I);
3   F(t) ← evaluate(P(t), μ);
4   while (ι(P(t), F(t), t_max) ≠ true) do
5       for i ← 1 to λ do
            P'(t) ← select(P(t), F(t));
            a'_i(t + 1) ← recombine(P'(t));
            a''_i(t + 1) ← mutate(a'_i(t + 1));
        od
6       P'(t + 1) ← (a''_1(t + 1), ..., a''_λ(t + 1));
7       F'(t + 1) ← evaluate(P'(t + 1), λ);
8       {P(t + 1), F(t + 1)} ← replace(P(t), P'(t + 1), F(t), F'(t + 1));
9       t ← t + 1;
    od
```

In order to implement a meta-evolutionary approach, the procedure *EA* is called as follows:

$$\text{Best\_EA} \leftarrow \text{EA}(\mu_\text{M}, \lambda_\text{M}, t_{\text{max\_M}}, F_\text{M}, I_\text{M}).$$

The result of this call, Best_EA, is the best EA ever found for the given base-level problem. The first three parameters of the procedure, μ_M, λ_M, and $t_{\text{max\_M}}$, denote the population size, offspring population size, and maximum number of generations for the metalevel EA, respectively. F_M is the fitness function for the metalevel problem, and I_M is the space of individuals $a_i(t)$ representing EA operator/parameter settings for the base-level problem.

The meta-evolutionary approach is realized by defining F_M to include a call to procedure *EA* as follows:

$$F_\text{M}(a_i(t)) = F_\text{B}(\text{EA}(\pi_{i_\mu}(a_i(t)), \pi_{i_\lambda}(a_i(t)), \pi_{i_{t_{\text{max}}}}(a_i(t)), F_\text{B}, I_\text{B})) \qquad \forall a_i(t) \in I_\text{M}$$

where:

- $\pi_{i_\mu}(a_i(t))$, $\pi_{i_\lambda}(a_i(t))$, and $\pi_{i_{t_{\text{max}}}}(a_i(t))$ denote the parent population size, offspring population size, and maximum number of generations for the base-level problem, respectively; they are obtained by applying the projection operator π_i to the current metalevel individual $a_i(t)$.
- F_B is the fitness function for the base-level problem.
- I_B is the space of individuals representing solutions to the base-level problem.

22.4 Parameter settings

In a meta-evolutionary approach, it is natural to ask how the types of operator and their parameter settings for the metalevel EA are determined. Several possibilities are described below.

(i) The operators and the parameter values of the metalevel EA are specified by the designer before the EA starts, and they do not change during the run. This approach is used by Bäck (1994), Grefenstette (1986), Lee and Takagi (1994), Mercer and Sampson (1978), Pham (1994), Shahookar and Mazumder (1990).

(ii) The operators and parameter values are initially determined by the designer and the metalevel EA is applied to find the best EA for a base-level problem which closely resembles the properties of the metalevel problem. Such a problem would require, for example, real-valued genes, the possibility to treat logical subgroups of genes as an atomic unit, and a sufficiently complex search space with multiple suboptimal peaks. The values obtained for the best base-level EA in this scenario are then copied and used as the parameter settings of the metalevel EA for optimizing the base-level EAs for the particular problem to be solved. This approach is used by Freisleben and Härtfelder (1993a).

(iii) The operators and the parameter values are initially determined by the designer and then copied from the best base-level EA for the problem in discussion (possibly after every metalevel generation). This, however, is only possible if the structural properties of the two problem types are identical. For example, it does not work if the base-level EA operates on strings for permutation problems, such as in combinatorial optimization problems like the traveling salesman problem (Freisleben and Merz 1996a, b, Goldberg and Lingle 1985, Oliver *et al* 1987). This approach is used by Freisleben and Härtfelder (1993b).

22.5 Theory

Some theoretical results which may assist in finding useful parameter settings for EAs have been reported in the literature. For example, there are theoretical investigations of the optimal population sizes (Goldberg 1989b, Nakano *et al* 1994, Ros 1989), the optimal mutation probabilities (Bäck 1993, De Jong 1975, Hesser and Männer 1990), the optimal crossover probabilities (Schaffer *et al* 1989), and the relationships between several evolutionary operators (Mühlenbein and Schlierkamp-Voosen 1995) with respect to simple function optimization problems.

However, it was shown by Hart and Belew (1991) that a universally valid optimal parametrization of a GA does not exist, because the optimal parameter values are strongly dependent on the particular optimization problem to be solved by the GA, its representation, the population model, and the operators used. The large number of possibilities precludes an exhaustive search of the space of operators and operator probabilities. This is a strong argument in favor of a (heuristic) meta-evolutionary or a self-*adaptive* approach.

22.6 Related work

Several meta-evolutionary approaches have been proposed in the literature, as discussed below.

(i) Mercer and Sampson (1978) gave presumably one of the first descriptions of a meta-evolutionary approach. The authors investigated the effects of two crossover and three mutation operators on a simple pattern recognition problem; the aim of their meta-algorithm was to determine the most suitable operator probabilities. They started with a population size of five individuals in the metalevel population and performed 75 generations, and their results were based on a single run (probably due to the limited computing power available at that time). Mercer and Sampson defined special types of metaoperator which were different from the genetic operators used in the base-level EAs. The parameter values of the base-level algorithm were adaptable over time: different operator probabilities were used in the different stages of the search.

(ii) Grefenstette's meta-GA (Grefenstette 1986) operated on individuals representing the population size, crossover probability, mutation rate, generation gap, scaling window, and selection strategy (i.e. proportional selection in pure and *elitist* form). The six-dimensional parameter space was discretized, and only a small number (8 or 16) of different values was permitted for each parameter. The base-level problems investigated were De Jong's set of test functions (De Jong 1975), and the control parameters of the meta-GA were simply set to those identified by De Jong in a number of experiments (population size $\lambda = 50$, crossover probability $p_c = 0.6$, mutation probability $p_m = 0.001$, a generation gap of 1.0, a scaling window of seven, elitist selection) (De Jong 1975). The meta-GA started with a population size of 50 and produced 20 generations. The results suggested the use of a significantly different crossover probability and a different selection scheme for the on-line and off-line performance of the GAs, respectively. In particular, the meta-evolution experiment yielded the result $\lambda = 30$ (80), $p_c = 0.95$ (0.45), $p_m = 0.01$ (0.01), a generation gap of 1.0 (0.9), a scaling window of one (one), elitist (pure) selection when the on-line (off-line) performance was used.

(iii) Shahookar and Mazumder (1990) used a meta-GA to optimize the crossover rate (for three different permutation crossover operators), the mutation rate, and the inversion rate of a GA to solve the standard cell placement problem for industrial circuits consisting of between 100 and 800 cells. Similar to Grefenstette's proposal, the individuals representing the GAs consisted of discrete (integer) values in a limited range. For the meta-GA, the population size was 20, the crossover probability 1.0, the mutation probability 0.2, and it was run for 100 generations. In the experiments it was observed that the GA parameter settings produced by the meta-GA approach need to examine 19–50 times fewer configurations as compared to a commercial cell placement system, and the runtimes were comparable.

(iv) Freisleben and Härtfelder (1993a, b) proposed a meta-GA approach which was based on a much larger space of up to 20 components, divided into *decisions* and *parameters*. A decision is numerically represented by the probability that a particular variant of an operator is selected among a limited number of variants available of that operator, and a parameter value is encoded as a real number associated with the selected variant. The authors performed two experiments, one in which the base-level GAs tried to solve a neural *network weight optimization problem* (strings with real-valued alleles) (Freisleben and Härtfelder 1993b) and another one where a set of differently complex symmetric traveling salesman problems (strings with integer-valued alleles) (Freisleben and Härtfelder 1993a) was solved. By providing means for investigating the significance of a decision for or against an operator variant on the solution process, and the significance of finding the optimal parameter value for a particular given operator variant, the authors demonstrated that making the right choice for some operator variants is more crucial than for others, and thus some light was shed on the relationships between the various GA features during the search process. For example, in the neural network weight optimization problem the operators which contributed most to the solution were the mutation value replacement method (adding a value to the old value to obtain the mutated value or overwriting the old value), the granularity of the crossover operator (applying crossover such that logical subgroups of genes stay together as a structural unit), the selection method, the distribution of the mutation function, and the decision for or against using the *elitist model*. The decision to apply the *crowding method* and the decision for mutating logical subgroups (i.e. the weights of a neuron) together became important after the 60th metalevel generation; that is, when other decisions and parameters had already been appropriately determined. In the experiment with the set of traveling salesman problems it was surprising to find that the type of permutation crossover operator) was far less significant than the choice of the mutation method or the use of the elitist model.

Furthermore, in this approach the parameter values of the base-level GAs are adaptable to the complexity of the problem instances. The results were based on 100 metalevel generations with a starting population size of 50. On the level of the meta-GA, the authors adopted the currently best base-level parameter settings dynamically after every metalevel generation (Freisleben and Härtfelder 1993b), and also used the parameter settings obtained in a previous meta-GA experiment (Freisleben and Härtfelder 1993a).

(vi) Bäck's meta-algorithm (1994) combines principles of ESs and GAs in order to optimize a population of GAs, each of which is determined by 10 (continuous and discrete) components (in binary representation) representing both particular operator variants and parameter values. The meta-algorithm utilizes concepts from ESs to mutate a subset of the components, while the remaining components are mutated as in genetic algorithms according to a uniform probability distribution over the set of possible values. The objective function to be optimized by the base-level GAs was a simple sphere model. The average final best objective function value from two independent runs served as the fitness function for the meta-algorithm. The GAs found in the metaevolution experiment were considerably faster than standard GAs, and theoretical results about optimal mutation rates were confirmed in the experiments.

(vii) Lee and Takagi (1994) have presented a meta-GA-based approach for studying the effects of a dynamically adaptive population size, crossover, and mutation rate on De Jong's set of test functions (De Jong 1975). The base-level GAs had the same components as the ones used by Grefenstette (1986), but they were additionally augmented by a genetic representation of a fuzzy system, consisting of a fuzzy inference engine and a fuzzy knowledge base, to incorporate a variety of heuristic parameter control strategies into a single framework. The parameter settings for the meta-GA were fixed; a population size of 10 was used, and the meta-GA was run for 100 generations. The results indicated that a dynamic mutation rate contributed most to the on-line and off-line performance of the base-level GAs, again confirming recent theoretical results on optimal mutation rates.

(viii) Pham (1994) repeated Grefenstette's approach for different base-level objective functions as a preliminary step towards a proposal called *competitive evolution*. In this method, several populations, each with different operator variants and parameter settings, are allowed to evolve simultaneously. At each stage, the populations' performances are compared, and only the best populations, i.e. the one with the fittest member and the one with the highest improvement rate in the recent past, are allowed to evolve for a few more steps, then another comparison is made between all the populations. The competitive evolution method, however, may not be regarded as a meta-evolutionary approach as described above; it is more a kind of parallel (island) population model with heterogeneous populations.

(ix) Tuson and Ross (1996a, b) have investigated the dynamic adaptation of GA operator settings by means of two different approaches. In their *coevolutionary* approach (Bäck 1996) (which is usually called self-adaptive) the operator settings are encoded into each individual of the GA population, and they are allowed to evolve as part of the solution process, without using a meta-GA. In the *learning rule adaptation* approach (Davis 1989, Julstrom 1995), information on operator performance is explicitly collected and used to adjust the operator probabilities. The effectiveness of both methods was investigated on a set of test problems. The results obtained with the coevolutionary approach on binary-encoded problems were disappointing, and the adaptation of operator settings by coevolution was found to be of little practical use. The results obtained using a learning-based approach were more promising, but still not satisfactory. The authors concluded that the adaptation mechanism should be separated from the main genetic algorithm and the information upon which decisions are made should explicitly be measured. This indicates that a meta-evolutionary approach for parameter adaptation promises to be superior to the approaches investigated by Tuson and Ross—in contrast to many publications on ESs and EP where the benefits of self-adaptation could be demonstrated (Bäck 1996).

22.7 Conclusions

In this section we have presented meta-evolutionary approaches to determine the optimal evolutionary algorithm (EA), i.e. the best types of evolutionary operator and their parameter settings, for a given problem. The basic idea of a meta-evolutionary approach is to consider the search for the best (base-level) EA as an optimization problem which is solved by a metalevel EA. We have presented an informal and formal description of the general meta-evolutionary approach, pseudocode for realizing it, some theoretical results, and related work done in the area.

The meta-evolutionary approaches proposed in the literature essentially differ in the way a base-level EA is encoded as a member of the metalevel population (depending on the base-level problem investigated and the number/type of operators and the parameter values to control them), the manner in which the types of operator and their parameter settings for the metalevel EA are determined, and the results obtained from the metaevolution experiments. A common feature is that meta-evolutionary approaches usually require a high amount of computation time, but fortunately it is quite straightforward to develop a parallel implementation based on a manager–worker scheme (Bäck 1994, Freisleben and Härtfelder 1993a, b).

References

Bäck T 1993 Optimal mutation rates in genetic search *Proc. 5th Int. Conf. on Genetic Algorithms (Urbana-Champaign, IL, July 1993)* ed S Forrest (San Mateo, CA: Morgan Kaufmann) pp 2–8

——1994 Parallel optimization of evolutionary algorithms *Parallel Problem solving from Nature—PPSN III (Proc. Int. Conf. on Evolutionary Computation and 3rd Conf. on Parallel Problem Solving from Nature, Jerusalem, October 1994) (Lecture Notes in Computer Science 866)* ed Y Davidor, H-P Schwefel and R Männer (Berlin: Springer) pp 418–27

——1996 *Evolutionary Algorithms in Theory and Practice* (New York: Oxford University Press)

Booker L 1987 Improving search in genetic algorithms *Genetic Algorithms and Simulated Annealing* ed L Davis (London: Pitman)

Davis L 1989 Adapting operator probabilities in genetic algorithms *Proc. 3rd Int. Conf. on Genetic Algorithms (Fairfax, VA, June 1989)* ed J D Schaffer (San Mateo, CA: Morgan Kaufmann) pp 61–9

——1991 *Handbook of Genetic Algorithms* (New York: Van Nostrand Reinhold)

De Jong K A 1975 *An Analysis of the Behaviour of a Class of Genetic Adaptive Systems* PhD Thesis, University of Michigan

Freisleben B and Härtfelder M 1993a In search of the best genetic algorithm for the traveling salesman problem *Proc. 9th Int. Conf. on Control Systems and Computer Science (Bucharest)* pp 485–93

——1993b Optimization of genetic algorithms by genetic algorithms *Proc. 1993 Int. Conf. on Artificial Neural Nets and Genetic Algorithms* ed R F Albrecht, C R Reeves and N C Steele (Vienna: Springer) pp 392–9

Freisleben B and Merz P 1996a A genetic local search algorithm for solving symmetric and asymmetric traveling salesman problems *Proc. 1996 IEEE Int. Conf. on Evolutionary Computation* (Piscataway, NJ: IEEE) pp 616–21

——1996b New genetic local search operators for the traveling salesman problem *Proc. 4th Int. Conf. on Parallel Problem Solving from Nature (Berlin, 1996) (Lecture Notes in Computer Science 1141)* ed H-M Voigt, W Ebeling, I Rechenberg and H-P Schwefel (Berlin: Springer) pp 890–9

Glover F 1990 Tabu search: a tutorial *Interfaces* **20** 74–94

Goldberg D E 1989a *Genetic Algorithms in Search, Optimization and Machine Learning* (Reading, MA: Addison-Wesley)

——1989b Sizing populations for serial and parallel genetic algorithms *Proc. 3rd Int. Conf. on Genetic Algorithms (Urbana-Champaign, IL, July 1993)* ed J D Schaffer (San Mateo, CA: Morgan Kaufmann) pp 70–9

Goldberg D E and Lingle R 1985 Alleles, loci and the travelling salesman problem *Proc. 1st Int. Conf. on Genetic Algorithms and their Applications (Pittsburgh, PA, July 1985)* ed J J Grefenstette (San Mateo, CA: Morgan Kaufmann) pp 154–9

Grefenstette J J 1986 Optimization of control parameters for genetic algorithms *IEEE Trans. Syst. Man Cybernet.* **SMC-16** 122–8

Hart W E and Belew R 1991 Optimizing an arbitrary function is hard for the genetic algorithm *Proc. 4th Int. Conf. on Genetic Algorithms (San Diego, CA, July 1991)* ed R K Belew and L B Booker (San Mateo, CA: Morgan Kaufmann) pp 190–5

Hesser J and Männer R 1990 Towards an optimal mutation probability for genetic algorithms *Proc. 1st Int. Conf. on Parallel Problem Solving from Nature (Dortmund, 1990)* (*Lecture Notes in Computer Science 496*) ed H-P Schwefel and R Männer (Berlin: Springer) pp 23–32

Jog P, Suh J Y and van Gucht D 1989 The effects of population size, heuristic crossover and local improvement on a genetic algorithm for the traveling salesman problem *Proc. 3rd Int. Conf. on Genetic Algorithms (Fairfax, VA, June 1989)* ed J D Schaffer (San Mateo, CA: Morgan Kaufmann) pp 110–5

Julstrom B A 1995 What have you done for me lately? Adapting operator probabilities in a steady-state genetic algorithm *Proc. 6th Int. Conf. on Genetic Algorithms (Pittsburgh, PA, July 1995)* ed L J Eshelman (San Mateo, CA: Morgan Kaufmann) pp 81–7

Lee M A and Takagi H 1994 A framework for studying the effects of dynamic crossover, mutation, and population sizing in genetic algorithms *Advances in Fuzzy Logic, Neural Networks and Genetic Algorithms* (*Lecture Notes in Artificial Intelligence 1011*) ed T Furuhashi (Berlin: Springer) pp 111–26

Mercer R E and Sampson J R 1978 Adaptive search using a reproductive meta-plan *Kybernetes* **7** 215–28

Michalewicz Z 1994 *Genetic Algorithms + Data Structures = Evolution Programs* (Berlin: Springer)

Mühlenbein H and Schlierkamp-Voosen D 1995 Analysis of selection, mutation and recombination in genetic algorithms *Evolution and Biocomputation* (*Lecture Notes in Computer Science 899*) ed W Banzhaf and F H Eeckman (Berlin: Springer) pp 142–68

Nakano R, Davidor Y and Yamada T 1994 Optimal population size under constant computation cost *Parallel Problem solving from Nature—PPSN III (Proc. Int. Conf. on Evolutionary Computation and 3rd Conf. on Parallel Problem Solving from Nature, Jerusalem, October 1994)* (*Lecture Notes in Computer Science 866*) ed Y Davidor, H-P Schwefel and R Männer (Berlin: Springer) pp 130–8

Oliver I M, Smith D J and Holland J R C 1987 A study of permutation crossover operators on the travelling salesman problem *Proc. 2nd Int. Conf. on Genetic Algorithms (Pittsburgh, PA, July 1987)* ed J J Grefenstette (San Mateo, CA: Morgan Kaufmann) pp 224–30

Ostermeier A, Gawelczyk A and Hansen N 1994 Step-size adaptation based on non-local use of selection information *Parallel Problem solving from Nature—PPSN III (Proc. Int. Conf. on Evolutionary Computation and 3rd Conf. on Parallel Problem Solving from Nature, Jerusalem, October 1994) (Lecture Notes in Computer Science 866)* ed Yu Davidor, H-P Schwefel and R Männer (Berlin: Springer) pp 189–98

Pham Q T 1994 Competitive evolution: a natural approach to operator selection *Progress in Evolutionary Computation (Lecture Notes in Artificial Intelligence 956)* ed X Yao (Berlin: Springer) pp 49–60

Reeves C R 1993 *Modern Heuristic Techniques for Combinatorial Problems* (New York: Halsted)

Ros H 1989 Some results on Boolean concept learning by genetic algorithms *Proc. 3rd Int. Conf. on Genetic Algorithms (Fairfax, VA, June 1989)* ed J D Schaffer (San Mateo, CA: Morgan Kaufmann) pp 28–33

Schaffer J D, Caruna R A, Eshelman L J and Das R 1989 A study of control parameters affecting online performance of genetic algorithms for function optimization *Proc. 3rd Int. Conf. on Genetic Algorithms (Fairfax, VA, June 1989)* ed J D Schaffer (San Mateo, CA: Morgan Kaufmann) pp 51–60

Shahookar K and Mazumder P 1990 A genetic approach to standard cell placement using meta-genetic parameter optimization *IEEE Trans. Computer-Aided Design* **CAD-9** 500–11

Starkweather T, McDaniel S, Mathias K, Whitley C and Whitley D 1991 A comparison of genetic sequencing operators *Proc. 4th Int. Conf. on Genetic Algorithms (San Diego, CA, July 1991)* ed R K Belew and L B Booker (San Mateo, CA: Morgan Kaufmann) pp 69–76

Tuson A L and Ross P 1996a Co-evolution of operator settings in genetic algorithms *Proc. 1996 AISB Workshop on Evolutionary Computing (Brighton, UK, April 1996)* ed T C Fogarty *(Lecture Notes in Computer Science 1143)* (New York: Springer) pp 120–8

——1996b Cost based operator rate adaptation: an investigation *Proc. 4th Int. Conf. on Parallel Problem Solving from Nature (Berlin, 1996) (Lecture Notes in Computer Science 1141)* ed H-M Voigt, W Ebeling, I Rechenberg and H-P Schwefel (Berlin: Springer) pp 461–9

van Laarhoven P J M and Aarts E H L 1987 *Simulated Annealing: Theory and Applications* (Boston, MA: Kluwer)

White T and Oppacher F 1994 Adaptive crossover using automata *Parallel Problem solving from Nature—PPSN III (Proc. Int. Conf. on Evolutionary Computation and 3rd Conf. on Parallel Problem Solving from Nature, Jerusalem, October 1994) (Lecture Notes in Computer Science 866)* ed Yu Davidor, H-P Schwefel and R Männer (Berlin: Springer) pp 1229–38

23

Coevolutionary algorithms

Jan Paredis

23.1 Introduction

The *Oxford Dictionary of Natural History* (Allaby 1985, p 150) provides the following description of coevolution:

> **coevolution** Complementary evolution of closely associated species. The interlocking adaptations of many flowering plants and their pollinating insects provide some striking examples of coevolution. In a broader sense, predator–prey relationships also involve coevolution, with an evolutionary advance in the predator, for instance, triggering an evolutionary response in the prey.

According to the description above, coevolution involves closely interacting species. In predator–prey systems, for example, there is an *inverse fitness interaction*: success on one side is felt by the other side as failure to which a response must be made in order to maintain one's chances of survival. There is a strong evolutionary pressure for prey to defend themselves better (e.g., by running faster or improved eyesight) in response to which predators will develop better attacking strategies (such as stronger claws or faster diving). This typically results in an *arms race* in which the complexity of both predator and prey increases stepwise.

This chapter describes the introduction of coevolution in EAs. The structure is as follows. First, a brief overview of research on competitive fitness functions is given. This work allows us to realize the (inverse) fitness interaction typical for competitive coevolution. The next section describes the research of Hillis (1992) on the computational use of predator–prey coevolution for solving a (nonbiological) optimization problem. Section 23.4 introduces a more recent CEA, called CGA, which has been applied to a large class of problems. Finally, a discussion is given in which various approaches are compared and some avenues for future research are given.

224

23.2 Competitive fitness

The fitness function of most EAs is a user-defined optimization criterion which evaluates the individuals in isolation from each other. This is not the case for *relative fitness functions*, which Angeline (Chapter 3, pp 1–2) defines as follows:

> Relative fitness measures access a solution's worth through direct comparison to some other solution either involved or provided as a component of the environment.

Competitive fitness functions (Angeline and Pollack 1993) are a type of relative fitness function. They calculate the fitness of an individual through competition 'duels' with other individuals. All that is required is to know which individual is better. No further quantification is needed.

Competitive fitness has been extensively applied to game playing. In this case, the individuals are game-playing strategies. The better individual in the duel is the strategy which wins the game. Besides these natural correspondences, there are several reasons for the use of competitive fitness in game playing. First, it is not safe to calculate the fitness of a strategy on the basis of its performance against a fixed set of strategies. This because there is a real danger that evolution will come up with a strategy which exploits specific weaknesses of the test strategies but which performs poorly against strategies not in the test set. It is the diversity in the population which ensures that the strategies are tested against a wide range of opponents. An example of this will be given shortly. Second, even if there is an optimal strategy which cannot be beaten by any other strategy, then that is the one the algorithm is actually searching for. Competitive fitness can be used when no such expert player is available. This because the self-scaling nature of competitive fitness allows for a gradual improvement of the quality of the playing strategies. In the beginning the strategies perform rather poorly, but as time passes the strategies improve. This, in its turn, places higher demands on the individuals because they face fiercer competition.

Applications using competitive fitness can be classified according to the type of pairwise *competition pattern* they use. All these interaction patterns strike a different balance between computational demand and the reliability of the fitness calculation. Some examples of competitive fitness functions operating on one population of interbreeding individuals are (i) 'all versus all' competition in which each individual is tested on each other individual (see, e.g., Axelrod 1987) (see later); (ii) random competition in which the fitness of an individual is determined by testing it on one (or a fixed number of) randomly chosen individual(s) (see, e.g., Reed *et al* 1967); (iii) tournament competition which uses a single-elimination, binary tournament to determine a relative fitness ranking (see, e.g., Angeline and Pollack 1993) (it is probably good to stress the difference from tournament selection, which uses absolute fitness values); (iv) 'all versus best' competition in which all individuals are tested with respect to the individual which is currently the fittest (see, e.g., Sims 1994).

The type of competition pattern is one criterion according to which applications using competitive fitness can be classified. Another is the *number of populations* involved. Does the competition occur between members of a single population or between members of two or more noninterbreeding populations? All patterns described above, except tournament selection, can easily be generalized to multispecies competition. Sims (1994) provides more detail on this issue.

As mentioned before, many of the applications using competitive fitness functions deal with game playing. The experiments of Reed *et al* (1967) with a simplified poker game were probably some of the first. More recently, competitive fitness was used for Tic Tac Toe (Angeline and Pollack 1993), Othello (Smith and Gray 1994), the Game of Tag (Koza 1992, Reynolds 1994), Backgammon (Pollack *et al* 1996), and for the evolution of a suitable behavior and morphology of virtual creatures competing with another for possession of a block initially placed in between them (Sims 1994).

Here, we concentrate on one of the best-known uses of a competitive fitness function, namely, Axelrod's (1984, 1987) experiments to evolve strategies for the *iterated prisoner's dilemma* (IPD). The IPD is based on the prisoner's dilemma, a simple two-person game that has been studied extensively in game theory. During such a game each of the players can choose one of the following actions: cooperation or defection. After both players have chosen their action each receives a (possibly different) payoff depending on its own action and the action chosen by its opponent. Table 23.1 depicts a typical payoff table: if both players cooperate (defect) they each receive a payoff of three (one). If one defects whilst the other cooperates then the defector gets a payoff of five, and the cooperator gets zero payoff.

Table 23.1. A payoff matrix for the prisoner's dilemma. The first number gives the payoff received by player A in the given situation.

		Player B	
		Cooperate	Defect
Player A	Cooperate	3, 3	0, 5
	Defect	5, 0	1, 1

Which strategy should an individual adopt in order to maximize its payoff? If it 'thinks' the other player will cooperate then defection is the best option (yielding a payoff of five instead of three). If, on the other hand, the other player defects then again defection is the best option (yielding a payoff of one instead of zero). Hence, in any case it is better to defect. If, however, the game is iterated (i.e., the two players play several—but an unknown number of—games in a row) and the outcome of earlier trials is taken into account then the players may cooperate.

In 1979 Axelrod organized two tournaments with strategies (i.e., computer programs) submitted by various researchers. These strategies determine their next action based on the three previous moves of the player itself and of its opponent. Axelrod's tournaments attracted 14 and 63 contestants, respectively. In both tournaments each program played against all other contestants (i.e., 'all versus all competition'). The same strategy won on both occasions. Even more surprisingly, this strategy is extremely simple: it cooperates in the first play; in all subsequent plays it chooses the same action as the other player did in the previous game. This strategy was called *tit for tat*.

As a next step, Axelrod (1984) used an EA to evolve a population of 20 players. Playing all 63 strategies would result in a computationally demanding fitness calculation. For this reason a fixed set of eight 'representative' players were selected. The fitness of a strategy was based on its performance against these eight strategies. In most of the experiments, the evolved strategies were similar to tit for tat. In some runs, however, the EA often found strategies with a substantially higher score. These strategies are, however, not 'better' than tit for tat. They are just reflective of the fixed environment of eight strategies used to evaluate the individuals. In other words, the EA evolves solutions which exploit characteristics of the environment but which not necessarily play well against other strategies.

In a second set of experiments, Axelrod (1987) did replace this static environment by using a competitive fitness in which each individual played against each of the other individuals in the population. Also in these experiments cooperation increased during a run of the EA, finally leading to variants of tit for tat. Since Axelrod's work the iterated prisoner's dilemma has become very popular in the EA community (see, for example, Fogel 1993).

23.3 Coevolving sorting networks

As far as the author is aware, the use of coevolution for computational purposes can be traced back to Barricelli (1962). He used simple objects (integers), which reproduced in a linear world. After some time, stable configurations (sequences) of numbers appeared. These configurations, called 'symbioorganisms', themselves displayed interesting behavior such as self-reproduction and crossover. Even phenomena like parasitism were observed. In a followup of this work, Barricelli (1963) let these symbioorganisms play a game of Tix Tax—which is similar to Nim—whenever they entered the same location. To this end, Barricelli implemented a procedure which translated the number sequences of the organisms into Tix Tax strategies. Probably due to the limited computer power available in those early days, only a few experiments were described. Despite large fluctuations in game-playing quality in different runs, Barricelli reported that the quality of play showed 'a tendency to increase' during a run.

More recently, Hillis (1992) put the computational use of predator–prey coevolution in the spotlight. He wanted to evolve networks for sorting lists of 16 numbers. Such a sorting network consists of comparisons between two numbers (and the possibility of swapping the two numbers). The goal is to find a network which correctly sorts all possible lists of 16 numbers with as few comparisons as possible.

Hillis used two populations. The first population consisted of sorting networks. The individuals in the second population were sets of test lists. These lists contain numbers to be sorted. Both populations were geographically distributed over a grid with each location containing one set of test lists and one sorting network. At each generation a sorting network was tested on the set of test lists at the same location. The fitness of a sorting network was defined as the percentage of correctly sorted test lists. The fitness of the set of test lists, on the other hand, was equal to the percentage of test lists incorrectly sorted by the network. This is similar to the inverse fitness interaction between predator and prey.

In both populations, selection is made locally: the higher the fitness of an individual in comparison with its neighbors, the higher its probability of being selected. Also reproduction occurs locally in both populations: individuals are recombined with individuals in their neighborhood.

Hillis has shown that, in comparison with traditional noncoevolutionary EAs, better sorting networks can be obtained. He mentions two reasons for this. First, both populations are in a constant flux. This prevents large portions of the population of sorting networks from becoming stuck in local optima. Second, fitness testing becomes much more efficient: one mainly focuses on test lists which the current population of sorting networks is unable to sort correctly. The particular competitive fitness pattern used in Hillis' system can be classified as *bipartite*. This because the fitness is calculated through the application of one sorting network on the set of test lists at the same location (Angeline and Pollack 1993).

23.4 A general coevolutionary genetic algorithm

This section describes a coevolutionary genetic algorithm (CGA) inspired by the work on sorting networks of Hillis. First, its operation is illustrated on an abstract class of problems. Next, some CGA applications are discussed.

23.4.1 Solving test–solution problems with a coevolutionary genetic algorithm

A large class of problems in artificial intelligence, computer science, and operations research involves the search for a solution which meets certain *a priori* criteria. The term 'test–solution problems' was introduced by Paredis (1996c) to refer to this class of problems. Inductive learning is an example of such a test–solution problem. It involves searching for an abstract concept description which generalizes given examples and excludes the given counterexamples. Constraint satisfaction is another example of the same type: one searches for a solution which satisfies the given constraints.

Test–solution problems clearly consist of two types of element: potential solutions (e.g., concept descriptions or potential solutions to a constraint satisfaction problem) and tests (e.g., examples or constraints). The basic population structures used by the CGA reflect this duality. In contrast with traditional single-population EAs, CGAs operate on two interacting populations: one consists of potential solutions; the other contains the tests.

Randomly generated solutions are used to initialize the solution population. The initial population of tests, on the other hand, contains the *a priori* given tests. The fitness of the initial individuals (tests and solutions) is calculated as follows. Each individual is paired up with 20 randomly chosen individuals of the other type. Hence, each initial solution is paired up 20 times with a randomly chosen test and vice versa. In other words, a solution encounters 20 tests. Each such *encounter* results in a fitness payoff for the solution of unity if it satisfies the test. Otherwise, it is zero. The fitness of the solution is now computed as the average payoff received by the solution. Furthermore, the payoffs are stored in a history associated with the solution. The fitness of the tests is calculated in a similar way: it is also based on the average payoff received during 20 encounters with randomly chosen solutions. The only difference is that there is an inverse fitness interaction between the two types of individual. Hence, the payoff received by a test is unity if it is violated by the solution it encounters. It is zero when the test is satisfied by the solution. As a result, solutions which satisfy more tests will—in general—be fitter. The inverse is true for tests: the fitter tests are violated more often by the solutions. Once the fitness of the initial populations is calculated, the individuals are sorted on fitness; that is, the fitter solutions and tests are located at the top of their respective populations.

Next, the basic cycle of a CGA is iterated. This cycle consists of the outermost 'for' loop in the pseudocode given below. First, 20 encounters take place. These encounters are similar to the ones described earlier, except that the function which SELECTs the individuals to be involved in an encounter is biased towards highly ranking individuals: the fittest individual is 1.5 times more likely to be selected than the individual with median fitness. Because of this linear rank-based selection, fit solutions and tests are often involved in an encounter. In other words, good solutions have to prove themselves more often. At the same time, the algorithm concentrates on satisfying difficult, that is, not yet solved, tests. This is because tests with a high fitness are violated by many members of the solution population. The function ENCOUNTER returns 1 if the solution satisfies the test. In the other case, 0 is returned. This is the payoff received by the solution. The function TOGGLE implements the inverse fitness interaction between tests and solutions. It changes a one to a zero and vice versa. Next, the histories of the solution and test involved in the encounter are UPDATEd. This is done by putting the payoff associated with the current encounter in the history of the individuals. At the same time the payoff associated with the least recent encounter is removed from these histories. Finally, the fitness of an individual is UPDATEd such that it is equal to the moving average of the payoffs it received during the 20 most recent encounters it was involved in. In order for the populations to remain sorted according to fitness, this change of fitness might move the individual up or down in its population.

```
{initialization of both populations}
SOL-POP ← CREATE-RANDOM-SOL-POP-INIT-FITNESS()
TEST-POP ← CREATE-RANDOM-TEST-POP-INIT-FITNESS()
FOR C ← 1 TO MAX-CYCLES DO
    FOR I ← 1 TO 20 DO
        {encounter}
        SOL ← SELECT(SOL-POP)
        TEST ← SELECT(TEST-POP)
        RES ← ENCOUNTER(SOL,TEST)
        UPDATE-HISTORY-AND-FITNESS(SOL,RES)
        UPDATE-HISTORY-AND-FITNESS(TEST,TOGGLE(RES))
    OD
    {solution reproduction}
    SOL1 ← SELECT(SOL-POP) {parent1}
    SOL2 ← SELECT(SOL-POP) {parent2}
    CHILD ← MUTATE-CROSSOVER(SOL1,SOL2)
    F ← FITNESS(CHILD)
    INSERT(CHILD,F,SOL-POP)
OD
```

After the execution of 20 encounters, a new solution individual is created. The same SELECTion mechanism as before is used to select two parents. Again, fitter solutions have a higher chance of being selected. Through the application of genetic operators (such as CROSSOVER and MUTATion) a new child is constructed. Its fitness is calculated as the average payoff received during 20 encounters with SELECTed tests. If this fitness is higher than the least-fit solution in the population, then the child is INSERTed in the population at the appropriate rank. At the same time, the least-fit solution is removed from the population. This one-at-a-time reproduction is common to all steady-state EAs, such as the $(\mu + 1)$ evolution strategies (Rechenberg 1973) or Genitor (Whitley 1989). Similar to the latter, our CGA uses adaptive mutation: the probability of mutating offspring becomes greater the more similar the parents are.

A couple of remarks are in order here. The continuous, partial fitness feedback resulting from the encounters is called *lifetime fitness evaluation* (LTFE). There are two parameters linked to the use of LTFE. The first is the size of the histories. The second is the number of encounters per cycle. Here, and in all other CGA applications, both parameters are set—quite arbitrarily—to 20. Other settings might further improve the results.

It should also be noted that the encounters used to calculate the fitness of a newly generated solution do not result in a fitness feedback for the tests involved. This in order to avoid unreliable fitness feedback resulting from possibly mediocre solution offspring.

Finally, in the particular context of test–solution problems, the tests are given beforehand. Hence, the population of tests contains the same individuals during the entire run. Only their ranking changes. Through the biased selection of tests involved in an encounter this indirectly changes the environment the solutions face. Some of the applications described below operate on two fully coevolving populations; that is, the test population evolves as well.

23.4.2 Coevolutionary genetic algorithm applications

Since 1994, the CGA has been used in a number of applications. These are briefly discussed here. For more detail and empirical results the interested reader is referred to the respective publications. Three of these applications evolve neural networks.

Classification. The oldest application (Paredis 1994a) evolves neural networks for classification. In this case, the given preclassified training examples form the test population. The solution population consists of feedforward neural networks with one hidden layer. During an encounter between a neural network and an example, the input nodes of the neural network are filled with values of the attributes of that example. Then propagation is performed. With each output node a class is associated. Hence, the number of output nodes is equal to the number of classes. After propagation the example's class is defined as the class associated with the output node with the highest activation value. The neural network receives a payoff of one (zero) if this class is (in)correct. Again, the examples get the 'inverse' payoff. Paredis (1994a) has shown that a CGA significantly outperforms traditional single-population GAs, in terms of solution quality as well as computation demand. Furthermore, this application clearly shows that the fitness proportional selection of examples (i.e. preclassified training examples) focuses the CGA on difficult—not yet solved—examples. From an early stage of the run, the fittest examples are those situated near the boundaries between different classes. These points are indeed hardest to classify: the training examples in their neighborhood belong to different classes.

Independent from the work described above, Siegel (1994) worked on a similar approach. He evolved decision trees and used fitness-based selection (using two-member tournament selection) of tests to calculate the fitness of a decision tree. His test population did not evolve either.

Process control. A second (neural network–CGA) application addresses process control (Paredis 1998). It focuses on a well-known bioreactor control problem which involves issues of delay, nonlinearity, and instability. The goal is to control the system such that a system variable becomes equal to a given value. Such control is to be achieved starting from points in a given area in the state space around the target point. Again, the solution population consists of feedforward neural networks (controllers) with one hidden layer. The test population consists of points in the designated area in the state space from where the controller should start. Now, also the tests evolve. The fitness payoff of a test (solution) is the (negative) temporal integral of the deviation from the target value. Again, the most difficult starting points—for example, those farthest away from the target point—rapidly become fit. During a CGA run the control consistently improves. Not only is the target value reached more quickly, later deviations around this point quickly decrease as well. This was observed for stable as well as unstable target points.

The work of Sebald and Schlenzig (1994) is relevant here. They evolved a population of controllers for highly uncertain plants. In fact, evolution was nested: in order to calculate the fitness of a controller, an EA was used to evolve plants which the controller had difficulty controlling. A CGA could be used to coevolve the controllers and the plants. Presumably, this requires far less computation than the nested EA of Sebald and Schlenzig. Rosin (1997) describes a coevolutionary approach to this same application, as well as some of the difficulties encountered.

Path planning. The third predator–prey application is somewhat related to the process control application. It involves a path-planning task in which neural networks have to navigate in a possibly cluttered world (Paredis and Westra 1997). Here again, there is a population of neural networks and a population of starting points on which the neural networks are tested.

Constraint satisfaction. In all three applications discussed above, the solution population consisted of neural networks. Obviously, a CGA can operate on various types of solution and test. Constraint satisfaction problems (CSPs) constitute an example of test–solution problems with considerably different types of individual. In a typical CSP, a solution consists of an array of variables which have to be assigned a value such that none of the *a priori* given constraints between the variables is violated. Here, the two coevolving populations consist of constraints and arrays of variables playing the role of tests and solutions, respectively. Paredis (1994b) shows that again, in comparison with traditional single-population GAs, the CGA finds high-quality solutions to CSPs in an efficient manner.

Density classification.   Another application coevolved *cellular automata* to classify the density of ones in bitstrings (Paredis 1997). Similar to the process control application, both the test population and the solution population really evolve. In this case, there is pressure on both populations to improve; that is, the solutions 'try' to correctly satisfy as many as possible individuals in the test population. At the same time the tests 'try' to make life hard for the solutions. It is not in their 'interest' that the problem is solved. Especially if there is no solution satisfying all possible tests then the tests might keep the solutions under constant attack. Even if there exists a solution satisfying all tests, it might be virtually unreachable given the coevolutionary dynamics. This might be caused by the high degree of epistasis in the linkages between the two populations. Due to such linkages a small change in the individuals in one population might require extensive changes in the other population. Paredis (1997) describes a first investigation into this matter. It provides a simple diversity-preserving scheme which prevents the tests from keeping ahead of the solutions. In fact, it achieves this by randomly inserting and deleting individuals from the test population. It also suggests other mechanisms, such as fitness sharing (see later) or the use of different reproduction rates in the two populations. The latter approach is currently being investigated.

Symbiosis.   All five applications above use predator–prey relations. Obviously, many other mechanisms—not necessarily based on inverse fitness interaction—exist in nature.   Symbiosis is such an important and widely occurring counterexample. It consists of a positive fitness feedback in which a success on one side improves the chances of survival of the other.

The only difference between a predator–prey CGA and its symbiotic variant is that due to the positive fitness interaction, the call to the function TOGGLE can be deleted from the code given earlier. Hence, both types of CGA clearly fit into one general framework. As far as we know, the use of symbiosis to speed up an EA is still largely unexplored. First experiments are reported by Paredis (1995, 1996a). They investigate the use of symbiosis to address the well-known EA coding problem of what the best 'genetic' representation of the individuals is. This choice of representation is often far from obvious. This because of the following contradiction. On the one hand, EAs are so-called weak-search methods, which are specifically suited for problems about which not much specific knowledge is available. On the other hand, an appropriate representation of the individuals is often vital in order for the EA to perform well. Unfortunately, it is often difficult—if not impossible—to find such a representation for problems about which very little is known.

A symbiotic CGA was used to coevolve solutions and their genetic representations (i.e., the ordering of the genes).   A representation adapted to the solutions currently in the population speeds up the search for even better solutions which in their turn might progress optimally when yet another representation is used. During the run of this symbiotic CGA the population of representations clearly groups functionally related good genes such that crossover can proliferate these groups of genes through the population. This evolutionary accumulation of improvements of the representation can be contrasted with the rather limited success obtained with the inversion operator. The random nature of inversion is probably the crux of the problem.

23.5 Discussion

The main purpose of the applications described above is to help us understand, explore, and illustrate the operation of CGAs. Recently, researchers have been using CGAs in real-world applications such as object motion estimation from video images (Dixon *et al* 1997) and timetabling for an emergency service (Kragelund 1997).

The research on CGAs demonstrates the generality and effectiveness of the combined use of LTFE and coevolution. The partial and continuous nature of LTFE is ideally suited to deal with coupled fitness landscapes. Coevolutionary interacting species typically give rise to such coupled landscapes: changes in one population have an impact on the fitness of the members of the other population.

Now, the CGA can be compared with other approaches. First of all, the fitness interaction in CGAs occurs between members belonging to different noninterbreeding populations. This is, for example, in contrast with Axelrod's (1987) experiments with the iterated prisoner's dilemma, which uses a single population. The same is true for the work of Angeline and Pollack (1993), Smith and Gray (1994), Reynolds (1994), and Pollack *et al* (1996). Koza (1992) used two species. Sims (1994) reports that he has used two-species as well as single-species runs. It should also be noted that CGAs aim at solving a wide range of problems. They are not restricted to the domain of game playing. Furthermore, the members of the CGA populations can be of different types (e.g., neural networks versus training examples). Only in the work of Hillis (1992) is this also the case.

The actual competition pattern of a CGA can be called *biased continuous* because of the bias in the encounters and the continuous nature of LTFE. This combination avoids the computational overhead of 'all versus all' testing. On the other hand, fit individuals are tested continuously during their lives. This concentrates the computational cost on promising solutions and discriminating tests. This also increases the reliability of the fitness calculation in comparison with, for example, random testing.

Hillis's work provided the basic inspiration for the research on CGAs. There are however some clear differences between the two. Through the introduction of LTFE, CGAs are finer grained and possibly also more robust. Moreover, various applications belonging to a large, abstract class of problems have demonstrated the power of CGAs. Whereas Hillis only used predator–prey interactions, it has been shown that other types of interaction (such as symbiosis) can easily be incorporated in CGAs as well.

More recently, Barbosa (1997) used predator–prey coevolution for structural optimization. Here, one population consists of mechanical structures. In the other population sets of external forces ('load sets') evolve. Fit mechanical structures are good at withstanding the load sets. Fit load sets, on the other hand, exploit specific weaknesses in the structures. This is a typical *minimax problem*. Barbosa's technique—as well as CGAs –can be applied to the large class of minimax problems. Hence, besides test–solution problems, minimax problems provide a second class of applications for predator–prey CGAs. Actually, the control problem of Sebald and Schlenzig (1994) described earlier is also a minimax problem.

The use of *fitness sharing* in a competitive coevolutionary context has already been investigated by a number of researchers. The Othello application of Smith and Gray (1994) is probably one of the earliest. Rosin and Belew (1995) combine competitive fitness sharing and shared sampling. The latter ensures that representative opponents are selected for the 'competition duels'. Juille and Pollack (1996) proposed another way to ensure population diversity: in a classification application the fitness is calculated by counting the number of cases a solution classifies correctly but which the solution it encounters wrongly classifies. Also of interest is the work by Darwen and Yao (1997). They introduce a fitness sharing method and test it on the iterated prisoner's dilemma. Interesting in their work is that, at the end of the run, they use the diversity in the population to select a strategy which plays best against specific unseen test opponents.

More recently, Rosin (1997) has observed that even with fitness sharing, solutions might 'forget' how to compete against old extinct types of individual which might be difficult to rediscover. For this reason he introduces a 'hall of fame', containing the best individuals of previous generations, which are used for testing new individuals. Rosin (1997) rightly observes that, similar to the hall of fame, the steady-state character of CGAs, together with the accumulation of fitness over long time scales, provides the CGA with memory. In order to keep the progress in the two populations balanced, Rosin (1997) introduces a 'phantom parasite' which provides a niche for 'interesting' individuals.

The work of Potter and De Jong (1994) on *cooperative coevolutionary genetic algorithms* (CCGAs) is relevant with respect to the symbiotic CGA. These CCGAs have been tested in the domain of function optimization. This way, the natural decomposition into a fixed number of subcomponents (namely the parameters of the function to be optimized) can be exploited. Each population contains values for one parameter. The fitness of a particular value is an estimate of how well it 'cooperates' with the other populations to produce a good overall solution. This involves global synchronous communication between the populations. The work on CCGAs and our work both use multiple interacting species to solve existing 'man-made' problems. The basic mechanisms are, however, quite different (e.g., the identity of the individuals and the fitness calculation). Moreover, CGAs do not require (knowledge about) the decomposition of solutions.

Finally, many avenues for future research into CEAs remain to be explored in order to fully understand and exploit the use of coevolution for various problems. Furthermore, the partial but continuous nature of LTFE allows for an elegant introduction of lifetime learning during the encounters. Paredis (1996b) describes some preliminary results on this subject. Also, more work is needed to investigate the conditions under which—and how—a CGA with fully evolving (predator–prey) populations can find good solutions.

References

Allaby M (ed) 1985 *The Oxford Dictionary of Natural History* (Oxford: Oxford University Press)

Angeline P J and Pollack J B 1993 Competitive environments evolve better solutions for complex tasks *Proc. 5th Int. Conf. on Genetic Algorithms (Urbana-Champaign, IL, 1993)* ed S Forrest (San Mateo, CA: Morgan Kaufmann) pp 264–70

Axelrod R 1984 *The Evolution of Cooperation* (New York: Basic)

——1987 Evolution of strategies in the iterated prisoner's dilemma *Genetic Algorithms and Simulated Annealing* ed L Davis (San Mateo, CA: Morgan Kaufmann) pp 32–41

Barbosa H J 1997 A coevolutionary genetic algorithm for a game approach to structural optimization *Proc. 7th Int. Conf. on Genetic Algorithms (East Lansing, MI, July 1997)* ed T Bäck (San Mateo, CA: Morgan Kaufmann) pp 545–52

Barricelli N A 1962 Numerical testing of evolution theories: part 1 *Acta Biotheor.* **16** 69–98

——1963 Numerical testing of evolution theories: part 2 *Acta Biotheor.* **16** 99–126

Darwen P J and Yao X 1997 Speciation as automatic categorical modularization *IEEE Trans. Evolutionary Comput.* **EC-1** 101–8

Dixon E L, Pantsios Markhauser C and Rao K R 1997 Object motion estimation technique for video images based on a genetic algorithm *IEEE Trans. Consumer Electron.*

Fogel D B 1993 Evolving behaviors in the iterated prisoner's dilemma *Evolutionary Comput.* **1** 77–97

Hillis W D 1992 Co-evolving parasites improve simulated evolution as an optimization procedure *Artificial Life II* ed C G Langton, C Taylor, J D Farmer and S Rasmussen (Redwood City, CA: Addison-Wesley) pp 313–24

Juille H and Pollack J B 1996 Co-evolving intertwined spirals *Evolutionary Programming V: Proc. 5th Ann. Conf. on Evolutionary Programming* (Cambridge, MA: MIT Press) pp 461–8

Koza J R 1992 *Genetic Programming* (Cambridge, MA: MIT Press)

Kragelund L V 1997 Solving a timetabling problem using hybrid genetic algorithms *Software—Practice and Experience* **27** 1121–34

Paredis J 1994a Steps towards co-evolutionary classification neural networks *Artificial Life IV* ed R Brooks and P Maes (Cambridge, MA: MIT Press) pp 102–8

——1994b Coevolutionary constraint satisfaction *Parallel Problem Solving from Nature—PPSN III (Proc. 3rd Int. Conf. on Evolutionary Computation and 3rd Conf. on Parallel Problem Solving from Nature, Jerusalem, October 1994) (Lecture Notes in Computer Science 866)* ed Yu Davidor, H-P Schwefel and R Männer (Berlin: Springer) pp 46–55

——1995 The symbiotic evolution of solutions and their representations *Proc. 6th Int. Conf. on Genetic Algorithms (Pittsburgh, PA, 1995)* ed L J Eshelman (San Mateo, CA: Morgan Kaufmann) pp 359–65

——1996a Symbiotic coevolution for epistatic problems *Proc. 12th Eur. Conf. on Artificial Intelligence (Budapest, August 1996)* ed W Wahlster (Chichester: Wiley) pp 228–32

——1996b Coevolutionary life-time learning *Parallel Problem Solving from Nature IV (Proc. 4th Int. Conf. on Parallel Problem Solving from Nature, Berlin, 1996) (Lecture Notes in Computer Science 1141)* ed H-M Voigt, M Ebeling, I Rechenberg and H-P Schwefel (Berlin: Springer) pp 72–9

——1996c Coevolutionary computation *Artific. Life J.* **2** 255–375

——1997 Coevolving cellular automata: be aware of the red queen! *Proc. 7th Int. Conf. on Genetic Algorithms (East Lansing, MI, July 1997)* ed T Bäck (San Mateo, CA: Morgan Kaufmann) pp 393–9

——1998 Coevolutionary process control *Proc. Artificial Neural Networks and Genetic Algorithms (Proc. Int. Conf., Norwich, April 1997)* ed G D Smith, N C Steele and R F Albrect (Vienna: Springer)

Paredis J and Westra R 1997 Coevolutionary computation for path planning *Proc. 5th Eur. Cong. on Intelligent Techniques on Soft Computing (EUFIT'97, Aachen, September 1997)* vol 1, ed H-J Zimmermann (Aachen: ELITE Foundation) pp 394–8

Pollack J B, Blair A D and Land M 1997 Coevolution of a backgammon player *Proc. 5th Artificial Life Conf. (Nara-shi, Japan, May 1996)* ed C Langton and K Shimohara (Cambridge, MA: MIT Press) pp 92–8

Potter M A and De Jong K A 1994 A cooperative coevolutionary approach to function optimization *Parallel Problem Solving from Nature—PPSN III (Proc. 3rd Int. Conf. on Evolutionary Computation and 3rd Conf. on Parallel Problem Solving from Nature, Jerusalem, October 1994) (Lecture Notes in Computer Science 866)* ed Yu Davidor, H-P Schwefel and R Männer (Berlin: Springer) pp 249–57

Rechenberg I 1973 *Evolutionstrategie: Optimiering technischer Systeme nach Prinzipien der biologischen Evolution* (Stuttgart: Frommann–Holzboog)

Reed J, Toombs R and Barricelli N A 1967 Simulation of biological evolution and machine learning *J. Theor. Biol.* **17** 319–42

Reynolds C W 1994 Competition, coevolution and the Game of Tag *Artificial Life IV* ed R Brooks and P Maes (Cambridge, MA: MIT Press) pp 59–69

Rosin C D 1997 Coevolutionary search among adversaries *PhD Thesis* University of California, San Diego, CA

Rosin C D and Belew R K 1995 Methods for competitive co-evolution: finding opponents worth beating *Proc. 6th Int. Conf. on Genetic Algorithms, ICGA 95 (Pittsburgh, PA, 1995)* ed L J Eshelman (San Mateo, CA: Morgan Kaufmann) pp 373–80

Sebald A V and Schlenzig J 1994 Minimax design of neural net controllers for highly uncertain plants *IEEE Trans. Neural Networks* **NN-5** 73–82

Siegel E V 1994 Competitively evolving decision trees against fixed training cases for natural language processing *Advances in Genetic Programming* ed K E Kinnear (Cambridge, MA: MIT Press) pp 409–23

Sims K 1994 Evolving 3D morphology and behavior by competition *Artificial Life IV* ed R Brooks and P Maes (Cambridge, MA: MIT Press) pp 28–39

Smith R E and Gray B 1994 Coadaptive genetic algorithms, an example in Othello strategies *TCGA Report* 94002, University of Alabama, Tuscaloga

Whitley D 1989 The Genitor algorithm and selection pressure: why rank-based allocation of reproductive trials is best *Proc. 3rd Int. Conf. on Genetic Algorithms (Fairfax, VA, 1989)* ed J D Schaffer (San Mateo, CA: Morgan Kaufmann) pp 116–23

24

Efficient implementation of algorithms

John Grefenstette

24.1 Introduction

The implementation of evolutionary algorithms requires the usual attention to software engineering principles and programming techniques. Given the variety of evolutionary techniques, it is not possible to present a complete discussion of implementation details. Consequently, this section focuses on a few areas that may substantially contribute to the overall efficiency of the implementation. These are (i) random number generators, (ii) genetic operators, (iii) selection, and (iv) the evaluation phase.

24.2 Random number generators

The subject of random number generation has a very extensive literature (Knuth 1969), so this section provides only a brief introduction to the topic. In fact, most programming language libraries include at least one random number generator, so it may not be necessary to implement one as part of an evolutionary algorithm. However, it is important to be aware of the properties of the random number generator being used, and to avoid the use of a poorly designed generator. For example, Booker (1987) points out that care must be taken when using simple multiplicative random number generators to initialize a population, because the values that are generated may not be randomly distributed in more than one dimension. Booker recommends generating random populations as usual, and then performing repeated crossover operations with uniform random pairing. Ideally, this would be done to the point of stochastic equilibrium, meaning that the probability of occurrence of every schema is equal to the product of the proportions of its defining alleles.

One commonly used method of generating a pseudorandom sequence is the *linear congruential method*, defined by the relation

$$X_{n+1} = (a_r X_n + c_r) \bmod m_r \qquad n \geq 0$$

where $m_r > 0$ is the *modulus*, X_0 is the starting value (or *seed*), and a_r and c_r are constants in the range $[0, m_r)$. The properties of the linear congruential method depend on the choices made for the constants a_r, c_r, and m_r. (See the book by Knuth (1969) for a thorough discussion.) Reasonably good results can be obtained with the following values on a 32-bit computer (Press *et al* 1988):

$$a_r = 4096$$
$$c_r = 150\,889$$
$$m_r = 714\,025.$$

The following routine uses the linear congruential method to generate a uniformly distributed random number in the range $[0, 1)$:

Input: the current random seed, *Seed*.
Output: u_r, a uniformly distributed random number in the range $[0, 1)$; the seed is updated as a side-effect.

1   $Urand(Seed)$:
2       $Seed \leftarrow (a_r\,Seed + c_r) \bmod m_r$;
3       $u_r \leftarrow Seed/m_r$;
4   **return** u_r ;

Given $Urand$ above, the following generates a uniformly distributed real value in the range $[a, b)$:

Input: lower bound a, upper bound b.
Output: u, a uniformly distributed random number in the range $[a, b)$.

1   $U(a, b)$:
2       $u_r \leftarrow Urand(Seed)$;
3       $u \leftarrow a + (b - a)\,u_r$;
4   **return** u ;

Note that the above procedure always returns a value strictly less than the upper bound b. The following generates an integer value from the range $[a, b]$ (inclusive of both endpoints):

Input: lower bound a (an integer), upper bound b (an integer).
Output: i, a uniformly distributed random integer in the range $[a, b]$.

```
1    Irand(a, b):
2        u_r ← Urand(Seed);
3        i ← a + ⌊(b − a + 1) u_r⌋;
4        return i ;
```

It is often required to generate variates from a normal distribution in evolutionary algorithms; for example, the mutation perturbations in evolution strategies (ESs) and evolutionary programming (EP) may be specified as normally distributed random variables. One way to implement an approximately normal distribution is based on the central-limit theorem, which states that, as $n \to \infty$, the sum of n independent, identically distributed (IID) random variables has approximately a normal distribution with a mean of $n\zeta$ and a variance of $n\sigma^2$, where ζ and σ^2 are the mean and variance, respectively, of the IID random variables. If the IID random variables follow the standard uniform distribution, as computed by $U(0, 1)$ above for example, then $\zeta = 1/2$ and $\sigma^2 = 1/12$. It follows that summing n samples from $U(0, 1)$ gives an approximation to a sample from a normal distribution with mean $n/2$ and variance $n/12$. The following illustrates this approach:

Input: mean ζ, standard deviation σ.
Output: x, a variate from the normal distribution with mean ζ and standard deviation σ.

```
1    N(ζ, σ):
2        sum ← 0;
         {n is a user-selected constant, n ≥ 12}
3        for i ← 1 to n do
4            sum ← sum + U(0, 1);
         od
5        z ← sum − n/2;
6        z ← z/(n/12);
         {z now approximates a sample from the standard normal
         distribution}
7        x ← ζ + σz;
8        return x ;
```

Studies have shown that fairly good results can be obtained with $n = 12$, thus eliminating the need for the division operation in the computation of the variance in line 6 (Graybeal and Pooch 1980).

Finally, evolutionary algorithms often require the generation of a randomized shuffle or permutation of objects. For example, it may be desired to shuffle a population before performing pairwise crossover. The following code implements a random shuffle:

Input: *perm*, an integer array of size *n*.
Output: *perm*, an array containing a random permutation of values from 1 to *n*.

```
1    Shuffle(perm):
2        for i ← 1 to n do
3            perm[i] ← i;
         od
4        for i ← 1 to n − 1 do
5            j ← Irand(i, n);
6            {swap items i and j}
7            temp ← perm[i] ;
8            perm[i] ← perm[j] ;
9            perm[j] ← temp;
         od
10       return perm ;
```

24.3 The selection operator

There is a wide variety of selection schemes for evolutionary algorithms. However, many selection algorithms involve two fundamental steps:

(i) Compute selection probabilities for the current population based on fitness.
(ii) Sample the current population based on the selection probabilities to obtain clones which may then be subject to mutation or recombination.

The SUS algorithm (Baker 1987) assigns a number of offspring to each individual in the population, based on the selection probability distribution. SUS is simple to implement and is optimally efficient, making a single pass over the individuals to assign all offspring.

24.4 The mutation operator

In contrast to ESs and EP, genetic algorithms usually apply mutation at a uniform rate (p_m) across all genes and across all individuals in the population. After the new population has been selected, each gene position is subjected to a Bernoulli trial, with a probability of success given by the mutation rate parameter p_m. The most obvious implementation involves sampling from a random variable for each gene position. However, if $p_m \ll 1$, the computational cost can be substantially reduced by avoiding the gene-by-gene calls to the random number generator. A sequence of Bernoulli trials with probability p_m has an interarrival time that follows a geometric distribution. A sample from such a geometric distribution, representing the number of gene positions until the next mutation, can be computed as follows (Knuth 1969, p 131):

$$m_{next} = \left\lceil \frac{\ln u}{\ln(1 - p_m)} \right\rceil$$

where u is a sample from a random variable uniformly distributed over $(0, 1)$. Assuming that the variable m_{next} is initialized according to this formula, the following pseudocode illustrates the mutation procedure for an individual:

Input: an individual a of length n; a mutation probability, p_m; the next arrival point for mutation, m_{next}.
Output: the mutated individual a_i, and the updated next mutation location, p_m.

1   mutate-individual(a, p_m, m_{next}):
        {Note: n is the number of genes in an individual}
2   **while** $m_{next} < n$ **do**
            {mutate-gene is a function that alters the individual a at }
            {position m_{next}, e.g. flip the bit at position m_{next} or sample}
            {from a distribution using $\sigma_{m_{next}}$, the standard deviation for}
            {this position.}
3   mutate-gene(a, m_{next});
            {compute new interarrival interval}
4   **if** $(p_m = 1)$
            then
5                $m_{next} \leftarrow m_{next} + 1$;
            else
6   sample $u \sim U(0, 1)$;
7                $m_{next} \leftarrow m_{next} + \left\lceil \dfrac{\ln u}{\ln(1 - p_m)} \right\rceil$;
            fi
        od
        {prepare for next individual, essentially treating the entire}
        {population as a single string of gene positions}
8        $m_{next} \leftarrow m_{next} - n$;
        return a, m_{next};

24.5   The evaluation phase

In practice, the time required to evaluate the individuals usually dominates the rest of the computational effort of an evolutionary algorithm. This is especially true if the computation of the objective function requires running a substantial program, such as a simulation of a complex system. It follows that very often the most significant way to speed up an evolutionary algorithm is to reduce the time spent in the evaluation phase. This section considers three ways to reduce evaluation time:

(i)   avoid redundant evaluations;
(ii) use sampling to perform approximate evaluations; and
(iii) exploit parallel processing.

24.5.1 Deterministic evaluations

The objective function may be deterministic (e.g. the tour length in a traveling salesman problem) or nondeterministic (e.g. the outcome of a stochastic simulation). If the objective function is deterministic, then it may be worthwhile to avoid evaluating the same individual twice. This can be implemented as follows:

(i)   When creating an offspring clone from a parent, copy the parent's objective function value into the offspring. Mark the offspring as *evaluated*.
(ii) After the application of a genetic operator, such as mutation or crossover, to an individual, check to see whether the resulting individual is different from the original one. If so, mark the individual as *unevaluated*.
(iii) During the evaluation phase, only evaluate individuals that are marked *unevaluated*.

The cost of this extra processing is dominated by step (ii), which can be accomplished in at most $O(n)$ steps for individuals of length n. The cost may be a constant for some operators (e.g. mutation). In many cases, this extra processing will be worthwhile to avoid the cost of an additional evaluation.

For some applications, it may be worthwhile to cache every individual along with its associated evaluation, so that the same individual is never evaluated twice during the course of the evolutionary algorithm. This method clearly involves significant additional overhead in terms of storage space and in matching each newly generated individual against the previously generated individuals, so its use should be reserved for applications with very expensive, but deterministic, objective functions.

24.5.2 Monte Carlo evaluation

If the objective function is nondeterministic, then each individual should be reevaluated during each generation. One important special class of nondeterministic objective functions is those computed through Monte Carlo procedures, in which a number of random samples are drawn from a distribution and the objective function returns the average of the sample values. If the objective function involves Monte Carlo sampling, then the user must determine how much effort should be expended per evaluation.

Fitzpatrick and Grefenstette (1988) discuss the case of a generational genetic algorithm using proportional selection, in which the evaluation of individuals is performed by a Monte Carlo procedure that iterates a fixed number of times, s, for each evaluation. It is shown that as the number of samples s is decreased (to save evaluation time), the accuracy of the estimation of a schema's fitness decreases much more slowly than the accuracy of the observed fitness of the individuals in the population. Assuming that the quality of the search performed by a genetic algorithm depends on the quality of its estimates of the performance of schemas, this suggests that genetic algorithms can be expected to perform well even using relatively small values for s. This analysis also suggests that the estimate of the average performance of the hyperplanes present in a given population may be improved by trading off an increase in the population size (thereby testing a greater number of representatives from each hyperplane) with a corresponding decrease in the number s of Monte Carlo samples per evaluation. The effect of this tradeoff on the overall runtime depends on the ratio of the evaluation costs to the other overhead associated with the genetic algorithm. A similar study by Hammel and Bäck (1994) showed a similar result, that ESs are also robust in the face of a noisy evaluation function. However, this study also showed that increasing the population size yielded a smaller performance improvement than increasing the sampling rate s for ESs.

24.5.3 Parallel evaluation

Since evolutionary algorithms are characterized by their use of a population, it is natural to view them as parallel algorithms. In generational evolutionary algorithms, substantial savings in elapsed time can often be obtained by performing evaluations in parallel. In the simplest form of parallelism, a master process performs all the function of the evolutionary algorithm except evaluation of individuals, which are performed in parallel by worker processes operating on separate processors. The master process waits for all workers to return the evaluated individuals before varying on with the next generation.

One definition of speedup for parallel algorithms is

$$S(p, N) = \frac{T(1, N)}{T(p, N)}$$

where $T(p, N)$ is the time required to perform a task of size N on p processors. The parallel evolutionary algorithm described above is a form of *parallel generate-and-test* algorithm, in which N possible solutions are generated on each iteration (i.e. in this context $N = \mu$, the population size). The efficiency of a parallel generate-and-test algorithm depends on ensuring that the workers finish their assigned tasks at nearly the same time. Since the master waits for all workers to complete, the time between the completion of a given worker and the completion of all others workers is wasted. For a such an algorithm, the maximum speedup with p processors is given by

$$S(p, N) = \frac{T(1, N)}{T(p, N)} = \frac{\alpha + \beta N}{\alpha + (\beta N / p)}$$

where α is the time required by the master process to generate the candidate solutions, and β is the time required to evaluate a single solution. Speedup approaches the ideal value of p as α/β approaches zero (Grefenstette 1995). In practice, the speedup is usually less than p due to communication overhead. Further degradation occurs if N is not evenly divided by p, since some workers will have more tasks to perform than others.

Other forms of parallel processing are discussed in Chapters 15 and 16.

References

Baker J E 1987 Reducing bias and inefficiency in the selection algorithm *Proc. 2nd Int. Conf. on Genetic Algorithms (Cambridge, MA, 1987)* ed J J Grefenstette (Hillsdale, NJ: Erlbaum) pp 14–21

Booker L B 1987 Improving search in genetic algorithms *Genetic Algorithms and Simulated Annealing* ed L Davis (San Mateo, CA: Morgan Kaufmann)

Fitzpatrick J M and Grefenstette J J 1988 Genetic algorithms in noisy environments *Machine Learning* **3** 101–20

Graybeal W J and Pooch U W 1980 *Simulation: Principles and Methods* (Cambridge, MA: Withrop)

Grefenstette J J 1995 Robot learning with parallel genetic algorithms on networked computers *Proc. 1995 Summer Computer Simulation Conf. (SCSC '95)* ed T Oren and L Birta (Ottawa: The Society for Computer Simulation) pp 352–57

Hammel U and Bäck T 1994 Evolution strategies on noisy functions: how to improve convergence properties *Parallel Problem solving from Nature—PPSN III (Proc. Int. Conf. on Evolutionary Computation and 3rd Conf. on Parallel Problem Solving from Nature, Jerusalem, October 1994)* (*Lecture Notes in Computer Science 866*) ed Yu Davidor, H-P Schwefel and R Männer (Berlin: Springer) pp 159–68

Knuth D E 1969 *Seminumerical Algorithms: the Art of Computer Programming* vol 2 (Reading, MA: Addison-Wesley)

Press W M, Flannery B P, Teukolsky S A and Vetterling W T 1988 *Numerical Recipes in C: the Art of Scientific Computing* (Cambridge: Cambridge University Press)

25

Computation time of evolutionary operators

Günter Rudolph and Jörg Ziegenhirt

25.1 Asymptotical notations

The computation time of the evolutionary operators will be presented in terms of asymptotics. The usual notation in this context is summarized below. Further information can be found, for example, in the book by Horowitz and Sahni (1978).

Let f and g be real-valued functions with domain $\mathbb{N}$.

(i) $f(n) = O(g(n))$ if there exist constants $c, n_0 > 0$ such that $|f(n)| \leq c\,|g(n)|$ for all $n \geq n_0$.
(ii) $f(n) = \Omega(g(n))$ if there exist constants $c, n_0 > 0$ such that $|g(n)| \leq c\,|f(n)|$ for all $n \geq n_0$.
(iii) $f(n) = \Theta(g(n))$ if $f(n) = O(g(n))$ and $f(n) = \Omega(g(n))$.
(iv) $f(n) = o(g(n))$ if $f(n)/g(n) \to 0$ as $n \to \infty$.

25.2 Computation time of selection operators

Suppose that the selection operator chooses μ individuals from λ individuals with $\mu \leq \lambda$. Thus, the input to the selection procedure consists of λ values $(v_1, \ldots, v_\lambda)$ usually representing the fitness values of the individuals. The output is an array of indices referring to the selected individuals. We shall assume that the fitness is to be maximized.

25.2.1 Proportional selection via roulette wheel

For proportional selection via the roulette wheel it is necessary to assume positive values v_k. At first we cumulate the values v_k such that $c_k = \sum_{i=1}^{k} v_i$ for $k = 1, \ldots, \lambda$. This requires $O(\lambda)$ time. The next step is repeated μ times: draw a uniformly distributed random number U from the range $(0, c_\lambda) \subset \mathbb{R}$. Since the values c_k are sorted by construction, binary search takes $O(\log \lambda)$ time to determine the index i with $c_i = \max\{k \in K : U < c_k\}$ where $K = \{1, \ldots, \lambda\}$. Consequently, the computation time is $O(\lambda + \mu \log \lambda)$ and $O(\lambda \log \lambda)$ if $\mu = \lambda$.

25.2.2   Stochastic universal sampling

The algorithm given by Baker (1987) is an almost deterministic variant of proportional selection as described above and requires $O(\lambda)$ time if $\mu = \lambda$.

25.2.3   q-ary tournament selection

Choose q indices $\{i_1, \ldots, i_q\}$ from $\{1, \ldots, \lambda\}$ at random and determine the index k with $v_k \geq v_i$ for all $i \in \{i_1, \ldots, i_q\}$. This is done in $O(q)$ time. Consequently, the μ-fold repetition of this operation requires $O(q\mu)$ time.

25.2.4 (μ, λ) selection

This type of selection can be done in $O(\lambda)$ time although all public domain evolutionary algorithm (EA) software packages we are aware of employ algorithms with worse worst-case running time. Moreover, it should be noted that this selection method can be interleaved with the process of generating new individuals. This reduces the memory requirements from λ to μ times an individual's size, which can be important when the individuals are very large objects or when computer memory is small. The FORTRAN code originating from 1975 (when memory was small) that comes with the disk of Schwefel (1995) realizes the following method. The first μ individuals that are generated are stored in an array and the worst one is determined. This requires $O(\mu)$ time. Each further individual is checked to establish whether its fitness value is better than the worst one in the array. If so, the individual in the array is replaced by the new one and the worst individual in the modified array is determined in $O(\mu)$ time. This can happen $\lambda - \mu$ times such that the worst-case runtime is $O(\mu + (\lambda - \mu)\mu)$. However there is a better algorithm that is based on *Heapsort*. After the first μ individuals have been stored in the array, the array is rearranged as a heap requiring $O(\mu)$ time. The worst individual in the heap is known by nature of the heap data structure. If a better individual is generated, the worst individual in the heap is replaced by the new one and the heap property is repaired which can be done in $O(\log \mu)$ time. Thus, the worst-case runtime improves to $O(\mu + (\lambda - \mu)\log \mu)$. For further details on heapsort or the heap data structure see for example the book by Sedgewick (1988).

Now we assume that λ individuals have been generated and the fitness values are stored in an array. The naive approach to select the μ best ones is to sort the array such that the μ best individuals can be extracted easily. This can be done in $O(\lambda \log \lambda)$ time by using, for example, heapsort. Again, there are better methods. Note that it is not necessary to sort the array completely—it suffices to create a heap in $O(\lambda)$ time, to select the best element, and to apply the heap repairing mechanism to the remaining elements. Since this must be done only μ times and since the repairing step requires $O(\log \lambda)$ time, the entire runtime is bounded by $O(\lambda + \mu \log \lambda)$. Other sorting algorithms can be modified for this purpose as well: a variant of *quicksort* is given by Press *et al* (1986), whereas Fischetti and Martello (1988) present a sophisticated quicksort-based method that extracts the μ best elements in $O(\lambda)$ time. Since the method is not very basic (and the FORTRAN code is optimized by hand!) we refrain from presenting a description here.

Table 25.1. Time and memory requirements of several methods to realize (μ, λ) selection.

No	Method	Worst-case runtime	Required memory
1	Schwefel (1975)	$O(\mu + (\lambda - \mu)\,\mu)$	$O(\mu)$
2	Schwefel + heap	$O(\mu + (\lambda - \mu)\log\mu)$	$O(\mu)$
3	Modified heapsort	$O(\lambda + \mu\log\lambda)$	$O(\lambda)$
4	Modified quicksort	$O(\mu\lambda)$	$O(\lambda)$
5	Fischetti–Martello	$O(\lambda)$	$O(\lambda)$

A summary of time and place requirements of the methods to realize (μ, λ) selection is given in table 25.1. We have made extensive tests to identify the break-even points of the above algorithms. Under the assumption that the fitness values in the array are arranged randomly and do not have a partial preordering, it turned out that up to $\lambda = 100$ and $\mu = 30$ all methods worked equally well. For larger $\lambda \in \{200, \ldots, 1000\}$ and $\mu/\lambda > 0.3$ method 2 clearly outperforms method 1, whereas methods 3–5 with $O(\lambda)$ place requirements do not reveal significant differences in performance, although method 4 has a trend to be slightly worse. Moreover, methods 2 and 4 perform similarly up to $\lambda = 1000$.

25.2.5 $(\mu + \lambda)$ selection

This type of selection can be realized similarly to (μ, λ) selection. If the generating and selecting process is interleaved, we need $O(\mu)$ place and $O(\lambda \log \mu)$ time. Assume that the old population of μ individuals is arranged in heap order and that a better individual was generated. Then only $O(\log \mu)$ time is necessary to replace the worst individual in the heap by the new one and to repair the heap, which can happen at most λ times. After all λ individuals are processed, the new population is of course in heap order. Therefore, once the initial population is arranged as a heap the population will remain in heap order forever.

If the λ individuals are generated before selection begins, we need $O(\mu + \lambda)$ place and $O(\mu + \lambda)$ time: simply run the algorithm of Fischetti–Martello on the array of size $\mu + \lambda$.

25.2.6 q-fold binary tournament selection (EP selection)

Similar to the selection methods given in subsections 25.2.1–25.2.3, this type of selection cannot be performed in an interleaved manner. The description given here follows Fogel (1995), p 137. Suppose that the λ individuals have been generated such that there is an array of $\mu + \lambda$ individuals. For each individual draw q individuals at random and determine the number of times the individual is better than the q randomly chosen ones. This number in the range 0–q is the score of the individual. Thus, the scores of all individuals can be obtained in $O((\mu + \lambda)q)$ time. Finally, the μ individuals with highest score are selected by the Fischetti–Martello algorithm in $O(\mu + \lambda)$ time. Alternatively, since the score can only attain $q + 1$ different values, the μ individuals with highest scores could be selected via bucket sort in $O(\mu + \lambda)$ time. Altogether, the runtime is bounded by $O((\mu + \lambda)q)$.

25.3 Computation time of mutation operators

We assume that the mutation operator is applied to an n-tuple and that this operation is an in-place operation. At first we presuppose that an elementary mutation of one component of the tuple can be done in constant time.

Let $p_i \in (0, 1]$ be the probability that component $i = 1, \ldots, n$ will undergo an elementary mutation. Then mutation works as follows: for each i draw a uniformly distributed random number $u \in (0, 1)$ and mutate component i if $u \le p$, otherwise leave it unaltered. Evidently, this requires $\Theta(n)$ time. If $p_i = 1$ for all $i = 1, \ldots, n$ we can of course refrain from drawing n random numbers. Although this does not decrease the asymptotic runtime, there will be a saving of real time. These savings can accumulate to a considerable amount of real time. Therefore, even apparently very simple operations deserve a careful consideration. For example, let $p_i = p \ll 1$ for all $i = 1, \ldots, n$ and let B be the number of components that will be affected by mutation. Since B is a binomially distributed random variable with expectation np, the average number of elementary mutations will reduce to 1 if $p = 1/n$. This can be realized by a simulation of the original mutation process. Imagine a concatenation of all n-tuples of a population of size λ such that we obtain a (λn)-tuple for each generation. If the EA is stopped after N generations the concatenation of all (λn)-tuples yields one large $(N\lambda n)$-tuple. Let U_k be a sequence of independent uniformly distributed random variables over $[0, 1)$. Then the random variable $M_k = 1_{[0,p)}(U_k)$ indicates whether component k of the $(N\lambda n)$-tuple has been mutated ($M_k = 1$) or not ($M_k = 0$). Let $T = \min\{k \ge 1 : M_k = 1\}$ be the random time of the first elementary mutation. Note that T has a geometrical distribution with probability distribution function $P\{T = k\} = p\,(1 - p)^{k-1}$ and expectation $E[T] = 1/p$. Since geometrical random numbers can be generated via

$$T = 1 + \left\lfloor \frac{\log(1 - U)}{\log(1 - p)} \right\rfloor$$

where U is a uniformly distributed random number over $[0, 1)$, we can simulate the original mutation process by drawing geometrical random numbers. Let T_ν denote the νth outcome of random variable T. Then the values of the partial sums of the series

$$\sum_{\nu=1}^{\infty} T_\nu$$

are just the indices of the $(N\lambda n)$-tuple at which elementary mutations occur. An implementation of this method (by drawing T on demand) yields a theoretical average speedup of $1/p$, but since the generation of a geometrical random number requires the logarithm function the practical average speedup is slightly smaller.

The initial assumption that elementary mutations require constant time is not always appropriate. For example, let $x \in X^n = \mathbb{R}^n$ and let $z \sim N(0, C)$ be a normally distributed random vector with zero mean and positive definite, symmetric covariance matrix C. Unless C is a diagonal matrix the components of the random vector z are correlated. Since we need $O(n)$ standard normally distributed random numbers and $O(n^2)$ elementary operations to build random vector z, the entire mutation operation $x + z$ requires $O(n^2)$ time; consequently, an elementary mutation operation requires $O(n)$ time.

25.4 Computation time of recombination operators

Assume that the input to recombination consists of $\rho \in \{2, \ldots, \mu\}$ n-tuples whereas the output is a single n-tuple. Consequently, for every recombination operator we have the bound $\Omega(n)$. Thus, the runtime for one-point crossover, uniform crossover, intermediate recombination, and gene pool recombination is $\Theta(n)$ whereas k-point crossover requires $O(n + k \log k)$ time.

Usual implementations of k-point crossover do not demand that the k crossover points are pairwise distinct. Therefore, we may draw k random numbers from the range 1 to $n - 1$ and sort them. These numbers are taken as the positions to swap between the tuples.

25.5 Final remarks

Without any doubt, it is always useful to employ the most efficient data structures and algorithms to realize variation and selection operators, but in almost all practical applications most time is spent during the calculation of the objective function value. Therefore, the realization of this operation ought to be always checked with regard to potential savings of computing time.

References

Baker J 1987 Reducing bias and inefficiency in the selection algorithm *Proc. 2nd Int. Conf. on Genetic Algorithms and their Applications (Pittsburg, PA, July, 1987)* ed Grefenstette J (Hillsdale, NJ: Erlbaum) pp 12–21

Fischetti M and Martello S 1988 A hybrid algorithm for finding the kth smallest of n elements in $O(n)$ time *Ann. Operations Res.* **13** 401–19

Fogel D B 1995 *Evolutionary Computation: Toward a New Philosophy of Machine Intelligence* (New York: IEEE)

Horowitz E and Sahni S 1978 *Fundamentals of Computer Algorithms* (London: Pitman)

Schwefel H-P 1995 *Evolution and Optimum Seeking* (New York: Wiley)

Sedgewick R 1988 *Algorithms* 2nd edn (Reading, MA: Addison-Wesley)

Press W H, Flannery B P, Teukolsky S A and Vetterling W T 1986 *Numerical Recipes* (Cambridge: Cambridge University Press)

26

Hardware realizations of evolutionary algorithms

Tetsuya Higuchi and Bernard Manderick

26.1 Introduction

In order to use evolutionary algorithms (EAs) including genetic algorithms (GAs) in real time or for hard real-world applications, their current speed has to be increased several orders of magnitude.

This section reviews research activities related to hardware realizations of EAs. First, we consider parallel implementations of GAs on different parallel machines. Then, we focus on more dedicated hardware systems for EAs. For example, a TSP GA machine, a wafer-scale GA machine, and vector processing of GA operators are described. Here, we discuss also a new research field, called *evolvable hardware* (EHW), since it has close relationships with hardware realizations of EAs.

This section is organized as follows. First, we discuss different PGAs. Second, we review dedicated hardware systems for EAs, and third, we take a closer look at EHW.

26.2 Parallel genetic algorithms

Parallel GAs (PGAs) can be classified along two dimensions. The first one is the parallel programming paradigm and the related computer architecture on which they are running. The second one is the structure of the population used. These two dimensions are not orthogonal: some population structures are better suited for certain architectures.

In the next three subsections, we discuss the different architectures and the different population structures and we classify PGAs according to these two dimensions.

26.2.1   Parallel computer architectures

Basically, there are two approaches to parallel programming. In *control level parallelism*, one tries to identify parts of an algorithm that can operate independently of each other and can therefore be executed in parallel. The main problem with this approach is to identify and synchronize these independent parts of the algorithm. Consequently, control level parallelism is limited in the number of parallel processes that can be coordinated. In practice, this limit lies at the order of ten.

In the second approach, *data level parallelism*, one tries to identify independent data elements that can be processed in parallel, but with the same instructions. It is clear that this approach works best on problems with large numbers of data. For these problems, data level parallelism is ideally suited to program *massively* parallel computers. Consequently, this approach makes it possible to fully exploit the power offered by parallel systems.

These parallel algorithms have to be implemented on one of the existing parallel architectures. A common classification of these is based on how they handle instruction and data streams. The two main classes are the *single instruction, multiple data* (SIMD) and the *multiple instruction, multiple data* (MIMD). An SIMD machine is executing a single instruction stream acting upon many data streams at the same time. Its advantage is that it is easily programmed. In contrast, an MIMD machine has multiple processors, each one executing independently its own instructing stream operating on its own data stream. MIMD computers can be divided into *shared-memory MIMD* and *distributed-memory MIMD* machines. They differ in the way the individual processors communicate. The processors in a shared-memory system communicate by reading from and writing to memory locations in a common address space. Since only one processor can have access to a shared memory location at a given time, this limits parallelization. Therefore, shared-memory systems are suited for control level parallelism but not for data level parallelism. However, these systems can be programmed easily. In a distributed-memory MIMD machine each processor has its own local memory. Communication between processors proceeds through passing data over a communication network. Many different network organizations are possible. The big advantage of distributed-memory MIMD machines is that they can be scaled to include a virtually unlimited number of processors without degrading performance. They are suited for both control level and data level parallelism. However, they are much more difficult to program.

26.2.2   Population structure in nature

Since the very beginning of population genetics, the impact of the population structure on evolution has been stressed: the way how a population is structured influences the evolutionary process (Wright 1931). A number of population structures have been introduced and investigated theoretically. We discuss them briefly. More details can be found elsewhere in this handbook.

The importance of this work for GAs is that it shows that PGAs are fundamentally different from the standard GA and that they different from each other depending on the population structure used.

According to Fisher, populations are effectively *panmictic* (Fisher 1930). This is, all individuals compete with each other during the selection process, and every individual can potentially mate with every other one. The standard GA has a panmictic population structure.

The island (Chapter 15) and the *stepping stone* population structures are closely related. In both cases, the population is divided into a number of *demes*, which are semi-independent subpopulations that remain loosely coupled to neighboring demes by migrants.

In general, the island models are characterized by relatively large demes, with all-to-all migration patterns between them. The stepping stone models are characterized by smaller demes arranged in a lattice, with migration patterns between nearest neighbors only.

Finally, in the *isolation-by-distance* model, the population is spread across a continuum. Each individual interacts only with individuals in its immediate neighborhood. Each small neighborhood is like a deme except that now the demes are overlapping. Individuals in a deme are implicitly isolated instead of explicitly as is the case in the two previous population structures.

26.2.3 An overview of parallel genetic algorithms

In this section, we give an overview of existing PGAs. For each PGA we discuss the *population structure* used, the amount of *parallelism*, and its *scalability*. Parallelism is measured by counting the number of individuals that can be treated in parallel during each step of the GA. Scalability reflects how an increase in population size affects the total execution time. These measures represent the two most important benefits of PGAs: the increased speed of execution and the possibility to work with large populations.

In *coarse-grained PGAs* the population is structured as in the stepping stone population model. A large population is divided into a number of equally sized subpopulations. The parallelism is obtained by the parallel execution of a number of classic GAs, each of which operates on one of the subpopulations. Occasionally, the parallel processes communicate to exchange migrating individuals. Because each process consists of running a complete GA, it is rather difficult to synchronize all processes. Consequently, coarse-grained PGAs are most efficiently implemented on MIMD machines. In particular, they are suited for implementation on distributed-memory MIMD systems where they can take full advantage of the virtually unlimited number of processors.

In coarse-grained PGAs, the parallelism is limited because each step still operates on a (sub)population. For instance, not all individuals can be evaluated simultaneously. The scalability of these PGAs is very good. If the population size increases, the total execution time can be kept constant by increasing the number of subpopulations. Moreover, a coarse-grained parallelization of GAs is the most straightforward way to efficiently parallelize GAs. Two early examples of coarse-grained parallelism are described by Tanse (1987) and Pettey *et al* (1987).

In *fine-grained PGAs* the population is structured as in the isolation-by-distance model. The population is mapped onto a grid and a neighborhood structure is defined for each individual on this grid. The selection and the crossover step are restricted to the individuals in a neighborhood. The parallelism is obtained by assigning a parallel process to each individual. Communication between the processes is only necessary during selection and crossover.

Fine-grained PGAs are ideally suited for implementation on SIMD machines because each (identical) process operates on only one individual, and the communication between processes can be synchronized easily. Implementation on MIMD machines is also easy. Another advantage is that there is no need to introduce new parameters and insertion and deletion strategies to control migration. The diffusion of individuals over the grid implicitly controls the migration between demes.

Fine-grained PGAs offer a maximal amount of parallelism. Each individual can be evaluated and mutated in parallel. Moreover, because the neighborhood sizes are typically very small, the communication overhead is reduced to a minimum. The scalability is also maximal because additional individuals only imply more parallel processes which do not affect the total execution time. Early examples of this approach are described by Gorges-Schleuter (1989), Manderick and Spiessens (1989), Mühlenbein (1989), Hillis (1991), Collins and Jefferson (1991), and Spiessens and Manderick (1991).

26.3 Dedicated hardware implementations for evolutionary algorithms

This section describes experimental hardware systems for GAs and classifier systems.

26.3.1 Genetic algorithm hardware

Four experimental systems have been proposed or implemented so far: the traveling salesman problem (TSP) GA engine at the University of Tokyo, the WSI- (Wafer-Scale-Integration-) based GA machine at Tsukuba University, the GA processor at Victoria University, and the GA engine in EHW at the Electrotechnical Laboratory (ETL).

The *TSP machine* (Ohshima *et al* 1995) has been developed to see whether or not common algorithms such as GAs can be directly implemented in hardware. The GA for TSP has been successfully implemented using Xilinx's FPGAs (field programmable gate arrays). The order representation for city tours is used to avoid illegal genotypes.

The 16 GA engines are implemented on one board. Two GA engines are developed on one FPGA chip, a Xilinx 4010 chip with maximum 10 000 gates. The GA engine implements a transformation followed by the fitness evaluation part. The transformation translates the order representation of a city tour into the path representation of that tour. In the engine, the transformation, the GA operations, and the fitness evaluations are executed in a pipeline.

The improvement in execution time compared with a SPARC station2 (50 MHz) is expected to be about 800-fold. On a SPARC, 25 generations take 9.1 s, on the eight FPGAs (20 MHz) running in parallel they take only 4.7 ms.

The *WSI-based GA machine* (Yasunaga 1994) developed at Tsukuba University consists of 48 chromosome chips on a 5 in wafer having in total 192 chromosome processors. The basic idea is as follows. It is inevitable to have electrically defective areas in a large wafer. However, the robustness of the GA may absorb such defects since it is expected to work even if defective chromosome processors emerge.

The chromosome chips are connected in a two-dimensional array on the wafer. On a chromosome chip, four chromosome processors are also connected in an array structure.

At Victoria University in Australia, a GA processor was designed and partially implemented with Xilinx LCA chips (Salami 1995). Its applications include adaptive IIR filters and PID controllers. The GA processor architecture is described in the hardware description language VHDL. From this description, the logic circuits for the processor are synthesized.

EHW, described in Section 26.4, is hardware which changes its own architecture using evolutionary computation. The EHW developed at the ETL is planned to include GA-dedicated hardware (Dirkx and Higuchi 1993) to cope with real-time applications. The key feature of this hardware is that GA operations such as crossover and mutation are executed *bitwise* and in parallel at each chromosome (Higuchi *et al* 1994b). So far, PGAs have not attempted the parallelization of the GA operators themselves.

26.3.2   Classifier system hardware

Although dedicated hardware systems for classifier systems have not yet appeared, the following four systems use associative memories or parallel machines in order to speed up the rule matching operations between the input message and the condition parts of the classifier rules.

Robertson's parallel classifier system, *CFS, was built on the Connection Machine (CM) which has an SIMD architecture with 64 000 1-bit processors. The speed of the system is independent of the number of classifiers (i.e. the execution speed is constant whenever the number of the rules is less than

65 000), but is dependent on the size of message list due to bit-serial processing algorithms of the CM (Robertson 1987).

Twardowski's learning classifier system is based on the associative memory architecture of Coherent Research Inc. The associative memory is used not only for the rule matching but also for other search operations such as parent selection (Twardowski 1994).

The GA-1 system is a parallel classifier system on the parallel associative processor IXM2 at ETL (Higuchi *et al* 1994a). IXM2, consisting of 73 transputers and a large associative memory (256 000 words), is used for rule matching. This is achieved one order of magnitude faster than on a Connection Machine-2 (Kitano *et al* 1991).

ALECSYS is a parallel software system, implemented on a network of transputers, that allows the development of learning agents with distributed control architecture (Dorigo 1995). An agent is modeled as a network of learning classifier systems (with the bucket brigade and a version of the GA) and is trained by reinforcement. ALECSYS has been applied to robot learning tasks in both simulated and real environments.

26.4 Evolvable hardware

26.4.1 Introduction

EHW is hardware which is built on FPGAs and whose architecture can be reconfigured by using evolutionary computing techniques to adapt to the new environment. If hardware errors occur or if new hardware functions are required, EHW can alter its own hardware structure in order to accommodate such changes.

Research on EHW was initiated independently in Japan and in Switzerland around 1992 (for recent overviews see Higuchi *et al* (1994b) and Marchal *et al* (1994), respectively). Since then, interest has been growing rapidly. For example, the first international workshop on EHW, *EVOLVE95*, was held in Lausanne, in October 1995.

Research on EHW can be roughly classified into two groups: engineering oriented and embryology oriented. The engineering-oriented approach aims at developing a machine which can change its own hardware structure. It also tries to develop a new methodology for hardware design: hardware design without human designers. This group includes activities in ETL, ATR HIP Research Laboratories, and Sussex University.

The embryology-oriented approach aims at developing a machine which can selfreproduce or repair itself. Research in the Swiss Federal Institute of Technology and ATR is along this approach. Both are based on two-dimensional cellular automata.

After a brief introduction of FPGAs and the basic idea of EHW, the current research activities on EHW will be described.

26.4.2 Field programmable gate arrays

An FPGA is a software-reconfigurable logic device whose hardware structure is determined by specifying a binary bitstring. The FPGA is the basis of EHW and it is described below.

The advantage of FPGAs is that only a short time is needed to realize a particular design or to change that design compared with ordinary gate arrays such as mask programmable gate arrays (MPGAs).

When some changes in the design are needed, the hardware description is revised using a textual hardware description language (HDL) and the new description is translated into a binary string. Then, that string is downloaded into the FPGA and the new hardware structure is instantaneously built on the FPGA. Since they are so easy to reconfigure, FPGAs are becoming very popular especially for prototyping.

The structure of an FPGA is shown in figure 26.1. It consists of *logic blocks* and *interconnections*. Each logic block can implement an arbitrary hardware function depending on the specified bitstring associated with that logic block. Another bitstring specifies which blocks can communicate over the interconnections. Thus, two bitstrings determine the hardware function of the FPGA and all these bits together are called the *architecture bits*.

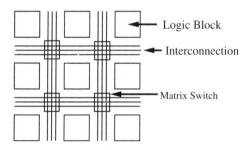

Figure 26.1. The architecture of an FPGA.

26.4.3 The basic idea behind evolvable hardware

In EHW, genotype representations of hardware functions are finally transformed into hardware structures on the FPGA. There are various types of representation.

For example, the basic idea of ETL's EHW is to regard the architecture bits of the FPGAs as genotypes which are manipulated by the GA. The GA searches for the most appropriate architecture bits. Once a good genotype is obtained, it is then directly mapped on the FPGA (Higuchi *et al* 1994b), as shown in figure 26.2.

The hardware evolution above is called *gate level evolution* because each gene may correspond to a primitive gate such as an AND gate.

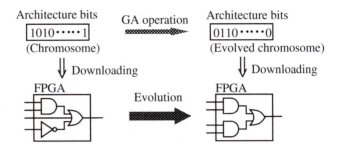

Figure 26.2. Hardware evolution by a GA.

26.4.4 The engineering-oriented approach

Three research activities along the engineering-oriented approach are described here.

Since 1992, ETL has conducted research on gate level evolution and developed two application systems. One is the prototypical welding robot of which the control part can be taken over by EHW when a hardware error occurs. By the GA, EHW learns the target circuit without any knowledge about the circuit while it is functioning correctly. The other is a flexible pattern recognition system which shares the robustness of artificial neural networks (ANNs) (i.e. noise immunity). While neural networks learn noise-insensitive functions by adjusting their weights and thresholds of neuron units, EHW implements such functions directly in hardware by genetic learning. Recently ETL initiated *function level evolution* where each gene corresponds to real functions such as floating multiplication and sine functions. Function level evolution can attain performance comparable to that of neural networks (e.g. two-spirals) (Murakawa *et al* 1996). For this evolution, a dedicated ASIC (application specific integrated circuit) chip is being developed.

Also, Thompson at the University of Sussex evolves at gate level a robot controller using a GA. For example, he evolved a 4 kHz oscillator and a finite-state machine that implements wall-avoidance behavior of a robot. The oscillator consisted of about 100 gates. The functions of the gates and their interconnections were determined by the GA (Thompson 1995).

Hemmi at ATR evolves the hardware description specified in the HDL by using genetic programming. The HDL used is SFL, which is a part of the LSI computer-aided design (CAD) system, PARTHENON. Therefore, once such a description is obtained, the real hardware can be manufactured by PARTHENON (Hemmi *et al* 1994).

Hardware evolution proceeds as follows. The grammar rules of SFL are defined first. The order of application of the grammar rules specifies a hardware description. Then, a binary tree is generated which represents the order of rule application. Genetic programming uses these trees then to evolve better ones.

So far, circuits such as adders have been successfully obtained.

26.4.5 The embryology-oriented approach

Ongoing research at the Swiss Federal Institute of Technology aims at developing an FPGA which can self-reproduce or self-repair (Marchal *et al* 1994).

A most interesting point is that the hardware description is represented by a binary decision diagram (BDD) and that these BDDs are treated as genomes by the GA. Each logical block of the FPGA reads the part of the genome which describes its function and is reconfigured accordingly. If some block is damaged, the genome can be used to perform self-repair: one of the spare logical blocks will be reconfigured according to the description of the damaged block.

Other research according to the embryology-oriented approach is de Garis' work at ATR. His goal is to evolve neural networks using a two-dimensional cellular automaton machine (MIT CAM8 machine) towards building an artificial brain. The neural network is formed as the trail on two-dimensional cellular automata by evolving the state transition rules by the GA (de Garis 1994).

26.5 Conclusion

As EA computation has inherent parallelisms, EA computation using parallel machines is a versatile and effective way for speeding up. However, the developments of dedicated hardware systems to EA computation are limited to experimental systems and this situation would not change drastically. This is because the development of dedicated systems may restrict careful tuning of parameters and strategies for selection and recombination, which affects EA performance considerably. If killer applications for EA are found, dedicated hardware systems will be developed more in the future.

Another new research area, EHW, has a strong potential to explore new challenging applications which have not been handled well so far due to the requirements of adaptive change and real-time response. These applications may include multimedia communications such as asynchronous transfer mode (ATM) and adaptive digital processing/communications. To be used in practice in such areas, current EHW needs to find faster learning algorithms and killer applications.

References

Collins R J and Jefferson D R 1991 AntFarm: towards simulated evolution *Artificial Life II* ed C G Langton, C Taylor, J D Farmer and S Rasmussen (Redwood, CA: Addison-Wesley) pp 579–602 de Garis H 1994 An artificial brain—ATR's CAM-brain project aims to build/evolve an artificial brain with a million neural net modules inside a trillion cell cellular atutomata machine *New Generation Computing* vol 12 (Berlin: Springer) pp 215–21

Dirkx E and Higuchi T 1993 *Genetic Algorithm Machine Architecture* Matsumae International Foundation 1993 Fellowship Research Report, pp 225–36

Dorigo M 1995 ALECSYS and the autonoMouse: learning to control a real robot by distributed classifier systems *Machine Learning* vol 19 (Amsterdam: Kluwer) pp 209–40

Dorigo M and Sirtori E 1991 Alecsys: a parallel laboratory for learning classifier systems *Proc. 4th Int. Conf. on Genetic Algorithms (San Diego, CA, July 1991)* ed R K Belew and L B Booker (San Mateo, CA: Morgan Kaufmann) pp 296–302

Fisher R A 1930 *The Genetical Theory of Natural Selection* (New York: Dover)

Gorges-Schleuter M 1989 ASPARAGOS: an asynchronous parallel genetic optimization strategy *Proc. 3rd Int. Conf. on Genetic Algorithms (Fairfax, VA, June 1989)* ed J D Schaffer (San Mateo, CA: Morgan Kaufmann) pp 422–7

Hemmi H, Mizoguchi J and Shimohara K 1994 Development and evolution of hardware behaviors *Proc. 4th Int. Workshop on the Synthesis and Simulation of Living Systems* ed R A Brooks and P Maes (Cambridge, MA: MIT Press) pp 371–6

Higuchi T, Handa K, Takahashi N, Furuya T, Iida H, Sumita E, Oi K and Kitano H 1994a The IXM2 parallel associative processor for AI *Computer* vol 27 (Los Alamitos, CA: IEEE Computer Society) pp 53–63

Higuchi T, Iba H and Manderick B 1994b Evolvable hardware *Massively Parallel Artificial Intelligence* ed H Kitano and J Hendler (Cambridge, MA: MIT Press) pp 398–421

Hillis W D 1991 Co-evolving parasites improve simulated evolution as an optimization procedure *Artificial Life II* ed C G Langton, C Taylor, J D Farmer and S Rasmussen (Redwood, CA: Addison-Wesley) pp 313–24

Kitano H, Smith S and Higuchi T 1991 GA-1: a parallel associative memory processor for rule learning with genetic algorithms *Proc. 4th Int. Conf. on Genetic Algorithms (San Diego, CA, July 1991)* ed R K Belew and L B Booker (San Mateo, CA: Morgan Kaufmann) pp 296–302

Manderick B and Spiessens P 1989 Fine-grained parallel genetic algorithms *Proc. 3rd Int. Conf. on Genetic Algorithms (Fairfax, VA, June 1989)* ed J D Schaffer (San Mateo, CA: Morgan Kaufmann) pp 428–33

Marchal P, Piguet C, Mange D, Stauffer A and Durand S 1994 Embryological development on silicon *Proc. 4th Int. Workshop on the Synthesis and Simulation of Living Systems* ed R A Brooks and P Maes (Cambridge, MA: MIT Press) pp 365–70

Mühlenbein H 1989 Parallel genetic algorithms, population genetics and combinatorial optimization *Proc. 3rd Int. Conf. on Genetic Algorithms (Fairfax, VA, June 1989)* ed J D Schaffer (San Mateo, CA: Morgan Kaufmann) pp 416–21

Murakawa M, Yoshizawa S, Kajitani I, Furuya T, Iwata M and Higuchi T 1996 Hardware evolution at function levels *Proc. 4th Int. Conf. on Parallel Problem Solving from Nature (Berlin, 1996) (Lecture Notes in Computer Science 1141)* ed H-M Voigt, W Ebeling, I Rechenberg and H-P Schwefel (Berlin: Springer) pp 62–71

Ohhigashi H and Higuchi T 1995 *Hardware Implementation of Computation in Genetic Algorithms* Electrotechnical Laboratory Technical Report

Ohshima R, Matsumoto N and Hiraki K 1995 Research on the reconfigurable engine for genetic computation *Proc. 3rd Japan. FPGA/PLD Design Conf.* (Tokyo: CMP Japan) pp 541–8 (in Japanese)

Pettey C B, Leuze M R and Grefenstette J J 1987 A parallel genetic algorithm *Proc. 2nd Int. Conf. on Genetic Algorithms (Cambridge, MA, 1987)* ed J J Grefenstette (Hillsdale, NJ: Erlbaum) pp 155–61

Robertson G 1987 Parallel implementation of genetic algorithms in a classifier systems *Genetic Algorithms and Simulated Annealing* ed L Davis (London: Pitman) pp 129–40

Salami M 1995 Genctic algorithm processor for adaptive IIR filters *Proc. IEEE Int. Conf. on Evolutionary Computing* (CD-ROM) (Casual) pp 423–8

Spiessens P and Manderick B 1991 A massively parallel genetic algorithm: implementation and first analysis *Proc. 4th Int. Conf. on Genetic Algorithms (San Diego, CA, July 1991)* ed R K Belew and L B Booker (San Mateo, CA: Morgan Kaufmann) pp 279–85

Tanese R 1987 Parallel genetic algorithms for a hypercube *Genetic Algorithms and their Applications: Proc. 2nd Int. Conf. on Genetic Algorithms (Cambridge, MA, 1987)* ed J J Grefenstette (Hillsdale, NJ: Erlbaum) pp 177–83

Thompson A 1995 Evolving electronic robot controllers that exploit hardware resources *Proc. 3rd Eur. Conf. on Artificial Life* (Berlin: Springer) pp 640–56

Twardowski K 1994 An associative architecture for genetic algorithm-based machine learning *Computer* vol 27 (Los Alamitos, CA: IEEE Computer Society) pp 27–38

Wright S 1931 Evolution in Mendelian populations *Genetics* vol 16

Yasunaga M 1994 Genetic algorithms implemented by wafer scale integration—wafer scale integration by LDA (leaving defects alone) approach *Trans. Inst. Electron. Information Commun. Eng.* **J77-D-I** 141–8 (Tokyo: Institute of Electronics, Information and Communication Engineers) (in Japanese)

Further reading

1. Gordon V S and Whitley D 1993 Serial and parallel genetic algorithms as function optimizers *Proc. 5th Int. Conf. on Genetic Algorithms (Urbana-Champaign, IL, July 1993)* ed S Forrest (San Mateo, CA: Morgan Kaufmann) pp 155–62

2. Spiessens P 1993 *Fine-Grained Parallel Genetic Algorithms: Analysis and Applications* PhD Thesis, AI Laboratory, Free University of Brussels

   These are good starting points for PGAs. Both give an overview of different PGAs and have extensive bibliographies.

3. Higuchi T, Iba H and Manderick B 1994b Evolvable hardware *Massively Parallel Artificial Intelligence* ed H Kitano and J Hendler (Cambridge, MA: MIT Press) pp 398–421

4. Marchal P, Piguet C, Mange D, Stauffer A and Durand S 1994 Embryological development on silicon *Proc. 4th Int. Workshop on the Synthesis and Simulation of Living Systems* ed R A Brooks and P Maes (Cambridge, MA: MIT Press) pp 365–70

   These articles provide recent overviews of EHW.

Index, Volume 2